Judith M. Matzka

Discrete Time Analysis of Multi-Server Queueing Systems in Material Handling and Service

AF286769

Wissenschaftliche Berichte des
Institutes für Fördertechnik und Logistiksysteme
des Karlsruher Instituts für Technologie
Band 75

Discrete Time Analysis of Multi-Server Queueing Systems in Material Handling and Service

by
Judith M. Matzka

Dissertation, Karlsruher Institut für Technologie
Fakultät für Maschinenbau, 2011
Referent: Prof. Dr.-Ing. K. Furmans
Korreferentin: Dr. M. Di Mascolo

Impressum

Karlsruher Institut für Technologie (KIT)
KIT Scientific Publishing
Straße am Forum 2
D-76131 Karlsruhe
www.ksp.kit.edu

KIT – Universität des Landes Baden-Württemberg und nationales
Forschungszentrum in der Helmholtz-Gemeinschaft

KIT Scientific Publishing 2011
Print on Demand

ISSN: 0171-2772
ISBN: 978-3-86644-688-5

Discrete Time Analysis of Multi-Server Queueing Systems in Material Handling and Service

Zur Erlangung des akademischen Grades eines

Doktors der Ingenieurwissenschaften

der Fakultät für Maschinenbau
des Karlsruher Instituts für Technologie (KIT)
genehmigte

Dissertation

von

Dipl.-Wi.-Ing. Judith M. Matzka

Tag der mündlichen Prüfung:	18. Mai 2011
Hauptreferent:	Prof. Dr.-Ing. K. Furmans
Korreferentin:	Dr. M. Di Mascolo

Vorwort

Die vorliegende Arbeit entstand während meiner Tätigkeit als wissenschaftliche Mitarbeiterin am Institut für Fördertechnik und Logistiksysteme des Karlsruher Instituts für Technologie (KIT). Dieses Vorwort möchte ich nutzen, um all denen zu danken, die zum Gelingen der vorliegenden Dissertationsschrift beigetragen haben.

Bei Herrn Prof. Dr.-Ing. Kai Furmans, Leiter des Instituts für Fördertechnik und Logistiksysteme, bedanke ich mich für die Übernahme des Hauptreferats und für das mir entgegengebrachte Vertrauen. Seine Forschungsarbeiten im Bereich der zeitdiskreten bedientheoretischen Methoden haben mich zu dieser Arbeit motiviert. Die mir gewährte Freiheit zum selbständigen Arbeiten hat maßgeblich zum Gelingen meiner Promotion beigetragen.

Frau Dr. Maria Di Mascolo vom Laboratoire G-SCOP in Grenoble danke ich für die Übernahme des Korreferats und besonders für die anregenden Diskussionen. Die daraus hervorgehenden Impulse haben die Arbeit stets bereichert.

Herrn Prof. Dr.-Ing. Carsten Proppe danke ich dafür, dass er den Vorsitz der mündlichen Prüfung übernommen hat.

Bedanken möchte ich mich auch beim Karlsruhe House of Young Scientists (KHYS) für die finanzielle Unterstützung durch ein Auslandsstipendium, das mir einen dreimonatigen Forschungsaufenthalt in Grenoble ermöglicht hat. In dieser Zeit hatte ich die Gelegenheit mich ausschließlich meiner Arbeit zu widmen, wodurch ich die analytisch schwierigsten Fragestellungen beantworten und meine Arbeit ein großes Stück vorantreiben konnte.

Den aktiven und ehemaligen Kollegen gilt mein herzlicher Dank für die anregenden Diskussionen und die konstruktive Kritik an meiner Arbeit. Die sehr angenehme Arbeitsatmosphäre und die zahlreichen gemeinsamen Unternehmungen werde ich stets in guter Erinnerung behalten.

Mein besonderer Dank gilt Christian Huber und Eda Özden, die mein Manuskript korrigiert haben.

Mein tiefster persönlicher Dank gilt meinen Eltern, die mich auf meinem Weg unterstützten. Meiner Schwester Manuela, meinen Freunden und Verwandten, und insbesondere meinem Verlobten Thomas Stoll, danke ich für den Rückhalt, den sie mir in den letzten Jahren gegeben haben und ohne den diese Arbeit vermutlich nicht entstanden wäre.

Karlsruhe, Mai 2011   Judith Matzka

Contents

Kurzfassung

Judith Matzka

Zeitdiskrete Analyse von Mehrkanalbediensystemen in Materialfluss und Service

Diese Arbeit beschäftigt sich mit der Entwicklung von analytischen Verfahren für die Leistungsbewertung von parallelen Bedienstationen in Materialflusssystemen.

Zur Leistungsbewertung von Materialflusssystemen werden, insbesondere in der Grobplanungsphase, häufig bedientheoretische Methoden verwendet, da mit deren Hilfe in kürzester Zeit eine Vielzahl verschiedener Planungsszenarien quantitativ untersucht und bewertet werden können. Während mit Hilfe der herkömmlichen zeitkontinuierlichen Bedientheorie eine Berechnung von Kennwerten nur in Form von Mittelwerten und Varianzen möglich ist, werden bei der zeitdiskreten Modellierung sämtliche Kenngrößen mit generellen Wahrscheinlichkeitsverteilungen beschrieben. Dadurch wird deren Aussagekraft entscheidend erhöht. Es wird dadurch möglich, Quantile von Kennwerten zu bestimmen und Materialflusssysteme so auszulegen, dass Kundenaufträge in einer vorgegebenen Zeit mit einer gewünschten Wahrscheinlichkeit erfüllt werden können.

Die parallele Bearbeitung von Aufträgen findet sich in logistischen Systemen in verschiedensten Formen, wie z. B. bei parallel eingesetzten Maschinen in der Produktion, bei parallel arbeitenden Kommissionierern in einem Distributionssystem, beim parallelen Einsatz von Gabelstaplern in einem Lager etc. Ein wichtiger Baustein zur Modellierung eines Materialflusssystems ist daher das Mehrkanalbediensystem, bei

dem mehreren parallelen Bedienstationen ein gemeinsamer Warteraum
vorgeschaltet ist.

In der vorliegenden Arbeit werden unter generellen stochastischen
Verteilungsannahmen Kennwerte eines Mehrkanalsystems bestimmt.
Für die Verteilung der Anzahl Kunden im System zum Ankunftszeit-
punkt eines beliebigen Kunden wird eine exakte Berechnungsmeth-
ode vorgestellt. Ebenso werden die Wartezeitverteilung sowie die
Durchlaufzeitverteilung exakt bestimmt. Für die Berechnung der
Verteilung der Zwischenabgangszeiten wird eine Approximationsmetho-
de vorgestellt.

Abstract

Judith Matzka

Discrete Time Analysis of Multi-Server Queueing Systems in Material Handling and Service

Scope of this doctoral thesis is the development of discrete time methods for the performance evaluation of parallel servers in material flow systems. For the analysis of material flow systems, especially in an early planning stage, queueing methods are well suited for the performance evaluation of many different scenarios in quite a short time. While queueing models in continuous time domain are calculating mean values and variances of performance figures, discrete time models are describing the performance measures by probability mass functions. Thus, the results are more significant, because quantiles of these distributions can be estimated. Therefore we are able to determine our ability to meet customer requests.

The parallel service of orders in material flow and service systems can be found in several different ways, like e. g. machines in a production system working in parallel, order pickers in a distribution system working in parallel, several forklifts in a warehouse, etc. An important element for modeling material flow and service systems is the multi-server queue, which contains multiple parallel servers with a shared waiting room. In this doctoral thesis, performance parameters of multi-server queueing systems are estimated under general stochastic assumptions. We present an exact calculation method for the discrete time distribution of the number of customers in the queueing system at the arrival moment of an arbitrary customer. The waiting time distribution and

3

the sojourn time distribution are estimated exactly, as well. For the calculation of the inter departure time distribution, we present an approximation method.

1. Introduction

Before you can score, you
must first have a goal.

Anonymous

In the recent years, the global competition has faced companies with the need for efficient processes and a quick and reliable response to customer demand. At the same time, the resource deployment has to be minimized to reduce costs. In a plant, inefficient processes cause waiting time which makes the major part of the sojourn time of orders. A reduction of these non-value adding time leads to shorter response times but also to decreasing inventory and thus less assets. So, the competitiveness of companies depends crucially on the performance of the material flow system. Robust and efficient processes are required already in the planning phase. Planners are asked to quickly develop a cost-efficient solution, where at the same time performance figures can be met.

The analysis of material and information flows in material handling and service systems is often realized via simulation because it is a very powerful tool with an enormous degree of freedom regarding modeling and level of detail. On the other hand, simulation is very time consuming and therefore expensive. It requires a lot of time for modeling, validation and performing experiments. Numerous simulation runs are required for a single experiment in order to achieve statistically reliable results. Especially in an early planning stage, when key figures for many different scenarios have to be calculated, analytical approaches are well-suited to support the planning of manufacturing systems. To build a model that faces the real world, we have to assure that stochastic events as demand, processing times, machine failures, scrap etc. are considered in an appropriate manner. In this respect, queueing theory proves to be a suitable analytical tool in the literature for stochastic modeling of such systems (see Arnold and Furmans 2009, Buzacott and Shanthikumar 1993 and Hopp and Spearman 2001). It enables planners to achieve an efficient operating point by comparing many different options in quite a short time.

1.1. Problem Description and Scope of the Thesis

With queueing models in the continuous time domain, average performance measures can be derived if some basic parameters have been determined from past observations of the system. However, performance measures based on system averages are not sufficient to verify whether the requested sojourn times of orders through the plant can be met with an acceptable probability, which usually lies between 95% and 99%, possibly depending on order types. Therefore, for the evaluation of design alternatives in respect to their ability to reach the requested sojourn time from order entry to exit, discrete time queueing models with generally distributed processes are proposed. Discrete time models calculate not only mean values of performance measures, but complete discrete time distributions, and thus, enable a more detailed description of the system behavior than analytical methods in continuous time domain. We are able to obtain quantiles of distributions, and can determine our ability to meet customer requests.

Discrete time queueing analysis also offers other advantages compared to continuous time analysis or simulation (see Schleyer and Furmans 2007b, Schleyer 2007 and Schleyer, Furmans and Di Mascolo 2007). In the discrete time domain, we are able to model stochastic processes exactly by using empirical data obtained by as-is analysis. The assumption that time is not continuous but discrete is not an essential restriction for modeling material flow and production systems. The travel time of a material hand-ling device, for example, can adopt only a few time values, which can be described very well with a discrete distribution. In contrast, modeling in the continuous time scale requires the existence of a theoretical distribution function or the description by their moments. The derivation of a theoretical function is time consuming and this function describes the real stochastic process with imprecision. Especially, the description of multi modal functions is difficult. Thus, by means of discrete time distributions a high degree of accuracy of modeling can be reached with a low effort in data acquisition.

Discrete time queueing analysis is also a highly accurate analytical tool for the interpretation of stochastic processes. In continuous time domain, the analysis of manufacturing systems by means of general queueing systems is based on the description of the stochastic processes by the first two moments. Using 2-parameter approximations can cause remarkable deviations from the exact solutions as shown by Schleyer (2007) and Schleyer and Furmans (2007a), because the results are not only influenced by the first two moments, but also by further moments like the skewness and the kurtosis. In contrast, the waiting time distribution of the discrete time G|G|1-queue estimated by the approach of Grassmann and Jain (1989), for example, is exact within an ϵ-neighborhood.

These advantages motivated the development of several new analytical approaches in the recent years. Starting from some basic network elements, dealing with a one-piece flow, many models for the handling of batches followed. Since there is still a lack of appropriate discrete time models for material flow processes, we are motivated to find new solutions for problems in this field. Especially the multi-server queueing model in discrete time domain with general distributed arrival and service processes (G|G|m queueing system) is an important model element, missing in the discrete time toolbox.

Parallel servers with one shared waiting room can be found in material flow systems in several different ways. Especially for the modeling of a job shop production, where groups of homogeneous machines are present, multi-server queueing models are required (see Furmans 2000). They are useful to dimension buffers in front of the machines and to get detailed information about the sojourn time of a certain job. But not only the material flow of production systems contains a lot of parallel servers. Also in distribution centers, we can find them from the receiving to the shipping area. Normally, there are parallel receiving doors with one yard, where trucks are unloaded. This is done by multiple fork lifts at the same time. The order-picking in a distribution center is processed by several workers simultaneously. The same situation is valid for the packing after the order-picking. The most important key figure for these systems is the sojourn time of customer orders. When we know the distribution of the sojourn time, we can determine the probability for an on-time order fulfillment and make a promise of delivery to our customer. An adequate model of our system also helps us to dimension the number of parallel servers (doors, fork lifts, staff,...), that are necessary to process the incoming goods. This is even more important, when we analyze a cross docking center, where goods are not stored and the sojourn times have to be very short. Thus, there is a necessity to have appropriate queueing models to analyze these systems. The applications of multi-server queueing models are numerous and some numerical examples can be found in chapter 7 of this thesis.

The scope of the current thesis is to propose analytical methods for the performance evaluation of multi-server queueing systems and to gain insights into the system's behavior. We will present exact calculation methods for the distribution of the number of customers at the arrival instant and the distribution of the waiting time. To be able to connect the multi-server queueing element to other nodes of a network, we present an approximation method for the distribution of the interdeparture time. Besides, we will show the applicability of these methods to problems in real material handling and service systems.

1.2. Organisation of the Thesis

A visual representation of the overall structure of the current work is given in figure 1.1. Chapters 1 to 3 intend to familiarize the reader with the fundamentals and the motivation of the thesis. After an introduction, we present basic definitions of probability theory focusing on the discrete time domain in chapter 2. In section 2.3, we present a model of the regarded $G|G|m$-queue with the corresponding input parameters and the output distributions. We give a summary of the existing discrete time queueing models in chapter 3. A literature review of relevant publications about multi-server queues is provided in the same chapter. The conclusion of our literature review leads to the necessity to develop new methods for the analysis of parallel servers in discrete time

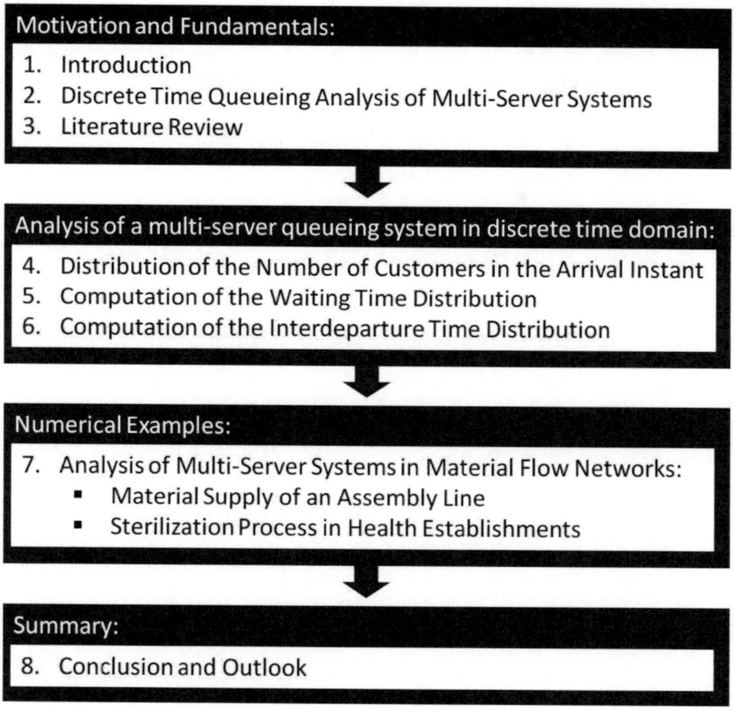

Motivation and Fundamentals:
1. Introduction
2. Discrete Time Queueing Analysis of Multi-Server Systems
3. Literature Review

Analysis of a multi-server queueing system in discrete time domain:
4. Distribution of the Number of Customers in the Arrival Instant
5. Computation of the Waiting Time Distribution
6. Computation of the Interdeparture Time Distribution

Numerical Examples:
7. Analysis of Multi-Server Systems in Material Flow Networks:
   - Material Supply of an Assembly Line
   - Sterilization Process in Health Establishments

Summary:
8. Conclusion and Outlook

Figure 1.1.: Organisation of the thesis

domain. In chapters 4 to 6, analytical methods for the determination of the output parameter distributions are presented. The distribution of the number of customers at the arrival instant is calculated in chapter 4. Based on the results of chapter 4, we determine the waiting time distribution in chapter 5. To link the multi-server queueing model to other nodes in a material flow network, we present an approximation method for the calculation of the interdeparture time distribution of the outgoing stream of customers in chapter 6. Chapter 7 illustrates how the presented methods can be used to model material flow networks. We present two numerical examples: a model of the material supply of an assembly line and a model of the sterilization process in health establishments, both with numerical results. We conclude this work in chapter 8, where the main results of this thesis are summarized and an outlook on further research is given.

2. Discrete Time Queueing Analysis of Multi-Server Systems

> In order to compose, all you
> need is to remember a tune
> that nobody else has thought
> of.
>
> *Robert Schumann*

The purpose of this chapter is to familiarize the reader with the fundamentals of discrete time queueing analysis. We are focusing on the discrete time domain and assume that the reader is familiar with the basics of probability and queueing theory. For a detailed introduction we refer to Kleinrock and Gail (1996) and Bolch (2006). For an introduction to the analysis of queueing systems in discrete time domain, we recommend the book of Tran-Gia (1996).

In section 2.1, we give a brief overview of basic definitions of probability theory in discrete time domain (compare Schleyer 2007), which we will use in our methods. We will also define the renewal process in the discrete time domain in section 2.2. After the definitions, we will describe the underlying discrete time queueing model for the calculation methods presented in this thesis (section 2.3).

2.1. Basic Definitions of Probability Theory in Discrete Time Domain

Analysis in discrete time domain assumes that time is not continuous but discrete. This means, that events are only recorded at discrete moments which are multiples of a constant increment t_{inc}. These events occur, when items are moved or when they change their status, for example by entering a queue, by being served, by merging with a stream of other items or at a split of a stream.

In our analysis, events are described by a discrete random variable. When we have given a discrete random variable X, we denote its distribution, which is also called probability mass function (pmf), by

$$P(X = i \cdot t_{inc}) = x_i \quad \forall i = 0, 1, ..., i_{max} \tag{2.1}$$

As a simplification we reduce this notation to

$$P(X = i) = x_i \quad \forall i = 0, 1, ..., i_{max} \tag{2.2}$$

P denotes a probability measure with a possible range of values from zero to a finite bound i_{max}. We can assume a finite value range i_{max} since this is in accordance with real applications. The probability vector of our pmf is denoted by

$$\vec{x} = \begin{pmatrix} x_0 \\ x_1 \\ \vdots \\ x_{i_{max}} \end{pmatrix} \tag{2.3}$$

The distribution function of X, which is called cumulative distribution function (CDF), is given by

$$P(X \leq i) = \sum_{j=0}^{i} x_j \quad \forall i = 0, 1, ..., i_{max}. \tag{2.4}$$

When we talk about a distribution in the subsequent chapters, we refer to the probability mass function. Several parameters, that are important for the analysis of stochastic systems, can be derived from the distribution of a discrete time random variable. We obtain the mean value of X by

$$E(X) = \sum_{i=0}^{i_{max}} i \cdot x_i. \tag{2.5}$$

The nth moment of X is defined as

$$E(X^n) = \sum_{i=0}^{i_{max}} i^n \cdot x_i. \tag{2.6}$$

The second central moment of our random variable X is also known as the variance and it is obtained by

$$VAR(X) = E(X^2) - E(X)^2. \tag{2.7}$$

The squared coefficient of variation (scv) is denoted by c_x^2. It is a normalized measure of statistical dispersion and is defined as

$$c_x^2 = \frac{VAR(X)}{E(X)^2}. \tag{2.8}$$

The scv is used to measure the process stability. Processes with low values of c_x^2 (near zero) indicate stable processes, and if c_x^2 is high, we have an unstable process. An exponential distributed process, for example, has an scv of $c_x^2 = 1$. The stability of a process has a direct impact on the number of customers waiting in a system and, thus, on the sojourn time of a customer through the system.

For the design of material flow systems it is interesting to know, if a process can be performed within a given time period with a given probability, e.g. the probability for an on-time order fulfillment. If the distribution of a process is known, this can be indicated by the appropriate quantile. The $u\%$-quantile of a discrete distribution is denoted by Q_u. It gives the value at which the CDF exceeds u percent and can be defined as

$$Q_u \Leftrightarrow P(X \leq Q_u) \geq u \wedge P(X \leq Q_u - 1) < u. \tag{2.9}$$

The distribution of the sum Z of two independent nonnegative random variables X and Y is called the convolution of their distributions and can be computed in the discrete case by

$$z_k = \sum_{i=0}^{k} x_i \cdot y_{j=k-i} \quad k = 0, 1, ..., k_{max} \quad with \; k_{max} = i_{max} + j_{max}.$$

(2.10)

The probability vector of the sum of the two random variables is given by

$$\vec{z} = \vec{x} \otimes \vec{y},$$
(2.11)

where $\otimes$ is defined as the convolution operator.

In our analysis, conditional probabilities are needed. Thus, we will give a short review. Let $P(A)$ denote the probability that the event A occurs, where $P(A)$ is a real number in the range of $0 \leq P(A) \leq 1$. The probability, that the event A occurs, under the condition that the event B has happened is denoted by $P(A|B)$ and is defined as

$$P(A|B) = \frac{P(A \cap B)}{P(B)}.$$
(2.12)

If a sequence of possible occurrences $\{A_i\}$ with their probability of appearance $P(A_i)$ is given, and we know the conditional probabilities $P(B|A_i)$, $P(B)$ can be calculated by the law of total probability:

$$P(B) = \sum_{i} P(B|A_i) \cdot P(A_i).$$
(2.13)

2.2. Discrete Time Renewal Process

Given is a sequence of events on a discrete time axis. We define a random variable X^n which describes the time interval between events n and $n - 1$. Its distribution is given by x_i^n, $i = 0, 1, 2, ..., i_{max}$. This sequence of events is defined as a discrete time renewal process if the length of all intervals is independent from each other and identically distributed. It follows that

$$x_i^n = x_i \ \forall \ n = 0, 1, 2, \dots \text{ and } \forall \ i = 0, 1, 2, \dots \tag{2.14}$$

When X^n is independent and identical distributed, each event marks a renewal point. The process is reset at the renewal point and the time interval to the next event is described by x_i. Observing a renewal process at an arbitrary time instant t^*, the time interval from t^* to the succeeding event is defined as the residual life time, denoted by R. The time interval from t^* to the preceding event is defined as the age of the process, denoted by U (see figure 2.1).

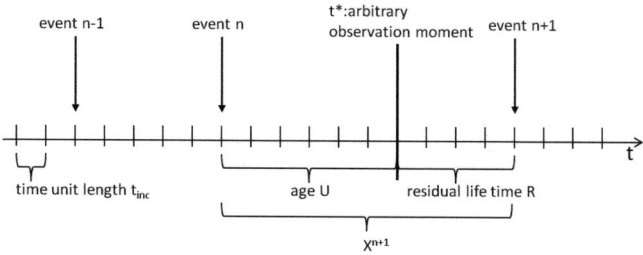

Figure 2.1.: Discrete time renewal process

Since time is assumed to be discrete, we have to distinguish if the arbitrary observation point t^* lies immediately before or immediately after discrete time instants (see Tran-Gia 1996). This leads to different distributions for the residual life time.

First, let us assume that t^* lies immediately before discrete time instants. If an event takes place at the observation instant, the occurrence of this event is observed and the residual life time is equal to zero. In this case, the age is the time period from the preceding event to t^*. In the second case, where we assume that t^* lies immediately after discrete time instants, the occurrence of this event is not observed and the residual life time is equal to the time interval until the occurrence of the succeeding event. In this case, the age is zero. According to these two different cases we can determine the range of possible residual life times of a given renewal process. If t^* lies immediately before discrete time instants, we conclude that the value range of R is $0 \leq R \leq x_{max} - 1$ and the value range of the age is $1 \leq U \leq x_{max}$. If t^* lies immediately after discrete time instants, R is defined from 1 to x_{max} and U from 0 to $x_{max} - 1$. For our analysis in the following chapters we define, that our observation points lie immediately before discrete time instants.

2.3. Model of a Multi-Server Queueing System in Discrete Time Domain

We will now define an adequate multi-server queueing model for the analysis of material flow systems. Figure 2.2 gives a visual representa-

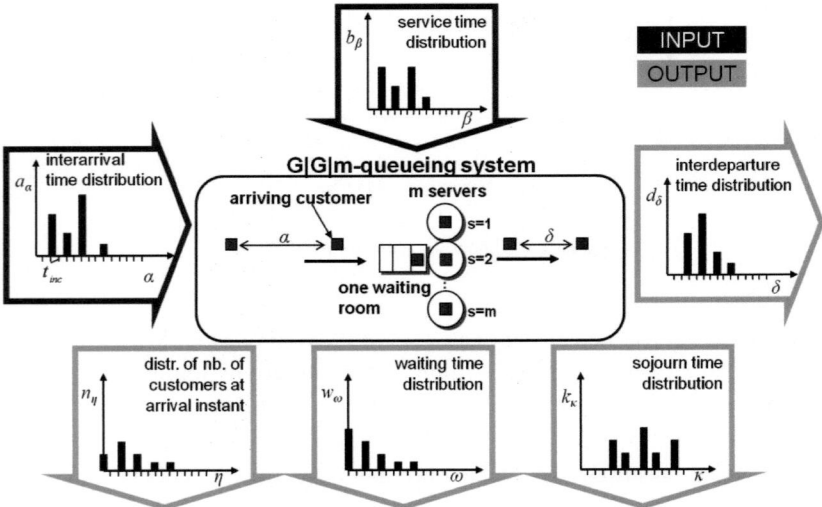

Figure 2.2.: Input and output parameters of a multi-server queueing model

tion of this model. The queueing system consists of m parallel servers and one shared waiting room. The customers arrive at the queueing system in single units and they are assigned to a free service station according to the order of their arrival (*First-Come, First-Served*). The time interval between the arrivals of two successive customers is called the interarrival time α and the probability of its appearance is given by the discrete probability distribution a_α. The service time β of a certain customer is given by the service time distribution b_β, which is the same for each of the m servers. In the moment when a customer arrives at the system, he finds a certain number of predecessors already present in the system. They are waiting or in process. The probability distribution of the number of customers, an arriving customer sees at his arrival instant, is defined as n_η. The number of customers in the system at the arrival instant decides, whether the arriving customer has to wait or not and thus influences the waiting time ω and its distribution w_ω. Knowing the waiting time distribution, we can determine the distribution of another important performance figure: the sojourn time κ. The notation of the sojourn time distribution is given by k_κ. To link the multi-server queue to other nodes of a network, we also have to know the interdeparture time distribution d_δ, where δ is the time interval between two successive departures.

3. Literature Review

> A person who walks in
> another's tracks leaves no
> footprints.
>
> *Pablo Picasso*

In the continuous time scale, an enormous amount of literature for queueing analysis exists. Some authors that provide a comprehensive insight and an overview about queueing theory, are for example Gnedenko and König (1983), Wolff (1989), Buzacott and Shanthikumar (1993), Kleinrock and Gail (1996) and Bolch (2006). That queueing theory in continuous time domain is well suited to model material flow systems is demonstrated in a variety of literature, such as Greiling (1997), Rall (1998) and Furmans (2000). In the continuous time domain, we can analyze general open queueing networks with the Queueing Network Analyzer of Whitt (1983), which consists of three basic models: the $G|G|1$-queue, the stochastic split and the stochastic merge. The analysis of manufacturing systems by means of general queueing systems in continuous time domain is based on two-parameter approximations. The first two moments of the interarrival and service time in a $G|G|1$-queue are used to calculate the mean waiting time and the first two moments of the interdeparture time using approximations. Using 2-parameter approximations can cause remarkable deviations from the exact solutions as shown by Schleyer (2007) and Schleyer and Furmans (2007a). Schleyer made some experiments in which he varied the discrete input distribution of his model, by keeping its mean value and variability constant. He showed that the results are not only influenced by these first two moments, but also by further moments like the skewness and the kurtosis. In contrast, the waiting time distribution of the discrete time $G|G|1$-queue estimated by the approach of Grassmann and Jain (1989), is exact within an ϵ-neighborhood. Thus, discrete time queueing analysis is a highly accurate analytical tool for the interpretation of stochastic processes. As mentioned before, discrete time analysis offers many advantages compared to continuous time modeling. The first discrete time queueing models were developed by telecommunication scientists in the 1980s to analyze the transmission of data packets in communication networks (see Ackroyd 1980, Walrand 1983, Hübner and Tran-Gia 1995, Tran-Gia 1996, Haßlinger 1995). These models allow the modeling of simple problems in material flow systems. Many authors, analyzing queueing systems in discrete time domain, are using geometric distributions to describe their processes (see e.g. Daduna 2001, Alfa 2002, Tran-Gia 1996, Haßlinger 1995). The geometric distribution is the discrete time counterpart of the exponential distribution, and thus has the *memoryless property*. This ends up in quasi-birth-and-death type Markov chains which have been well researched over the years. Most of the solutions in this area are based on the so-called matrix-geometric method (see Neuts 1981). Geometric distributions are well suited to model telecommunication networks. However, for many problems in material flow systems the geometric distribution is not applicable. Therefore, the approximation of data from an as-is analysis by a geometric distribution can cause significant errors which can be avoided by directly using the empirical data. Thus, we prefer using general discrete time queueing methods, that are dealing with the arbitrary general distributions obtained e.g. from the evaluation of a material flow system.

In section 3.1 we give an overview of concerning queueing models in the discrete time domain dealing with generally distributed processes. Starting from some basic elements of modeling, many methods have been developed for the analysis of batch processes in discrete time domain. Exact calculation methods for multi-server queues with discrete *and* general arrival and service processes do not exist so far. An overview of existing methods for the performance analysis of multi-server queues with generally distributed arrival and service processes in continuous time domain is given in section 3.2.

3.1. Discrete Time Methods

As for the Queueing Network Analyzer in the general continuous time domain, we have three basic model elements in the general discrete time "toolbox" as well:

- the G|G|1-queue with a discrete time distribution of interarrival and service time
- the split operation in order to split an incoming stochastic stream into two or more outgoing streams
- the merge of independent stochastic streams

All nodes of a network are treated as stochastically independent. In closed queueing networks with a constant number of customers, the system states of the different nodes are not independent from each other. Thus, the methods presented in this chapter can only be used to model open queueing networks. In order to connect the nodes, the departure process of each of the named model elements has to be known. Furthermore, the waiting times of the G|G|1-queues have to be determined, to be able to calculate the sojourn time distribution. Grassmann and Jain (1989) present a fast numerical method for the calculation of the waiting time distribution which is based on a Wiener-Hopf factorization of the underlying random walk. The authors also suggest an approach for the calculation of the interdeparture time distribution which is assumed to be the arrival stream for the succeeding node (see Jain and Grassmann 1988). Furmans and Zillus (1996) analyze the distribution of the number of customers at the arrival instant in a G|G|1-queue. This method can be used for the sizing of a buffer in front of a machine in order to guarantee a certain capacity. Furmans (2004) presents analytical methods for the computation of the interdeparture time distributions of the stochastic split and merge operation for a one-piece flow in discrete time domain. A visual representation of the before mentioned model elements is given in figure 3.1. The three named basic operations allow a rough and a fast analysis of material flow systems. To increase the level of detail, several models dealing with batch processes were developed in the recent years.

In material flow systems, many operations are done in batches. Batches are used for transports, because it is more economic to handle several

Element	Input	Output	Ex./ap.	author
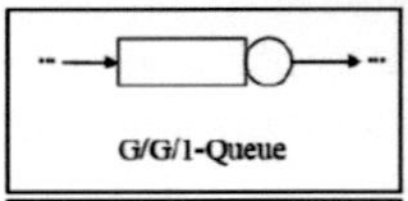 G/G/1-Queue	arrival, service	waiting, departure	exact	Grassman/ Jain 1988/1989
		number of customers	exact	Furmans/ Zillus 1996
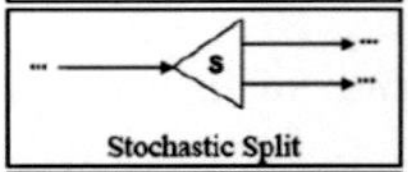 Stochastic Split	arrival, relation of directions	departure1, departure2	exact	Furmans 2004
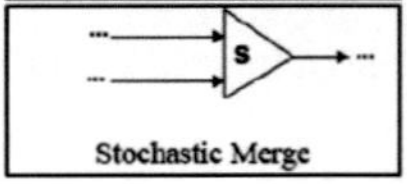 Stochastic Merge	arrival1, arrival2,	departure	approx. good for high variabilities	Furmans 2004

Figure 3.1.: Basic elements for the analysis of material flow systems

material units within one transport unit (e.g. pallet, plastic bin). If in some process steps we handle batches, there must have been a batch building process earlier in the network. Figure 3.2 gives an overview of existing batch building elements. It can be distinguished between two basic batch building modes: Collecting of orders until a predetermined amount of orders is reached or until a predetermined collecting time interval is elapsed (see Schleyer 2007). In addition, there are possible modifications of these basic batch building modes. An example is the minimum batch size rule, in which the collecting time is at least t_{out} time units. When t_{out} ends and less than a minimum amount of L orders were collected, the batch building process continues until the required number of L customers is attained. For the named batch building modes, Schleyer (2007) developed exact mathematical methods for the calculation of the waiting and interdeparture time distribution and the distribution of the departing batch size under general assumptions of the discrete input distributions. Özden et al. (2010) extended this toolbox by an element for a batch building mode, where orders are collected until a predetermined collecting time interval is elapsed, or a maximum batch size (e.g. container size) is reached.

The collected batches can then arrive at a service station, e.g. a machine, that serves single customers. Schleyer and Furmans (2007a) introduce a method for the determination of the waiting time distribution of the discrete time G|G|1-queue with batch arrivals and service of sin-

Element	Input	Output	Ex./ap.	author
Capacity rule 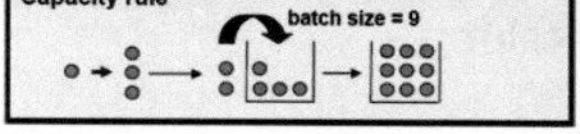 — batch size = 9	arrival time, batch size, capacity K	waiting time, departure time	exact	Schleyer 2007
Timeout rule 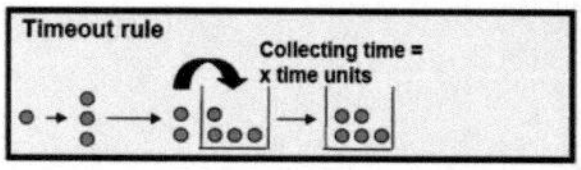 — Collecting time = x time units	arrival time, batch size, collecting time t_{out}	batch size, waiting time	exact	Schleyer 2007
Minimum batch size rule 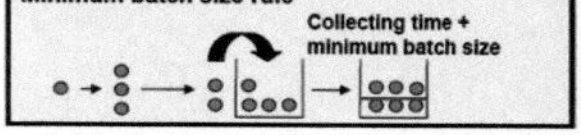 — Collecting time + minimum batch size	arrival time, batch size, collecting time t_{out}, minimum batch size L	batch size, waiting time, departure time	exact	Schleyer 2007
Capacitated timeout rule 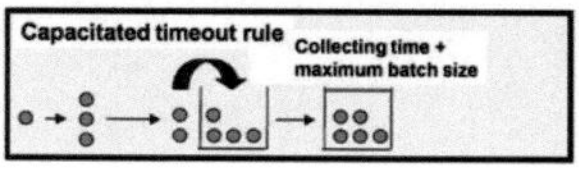 — Collecting time + maximum batch size	arrival time, batch size, collecting time, maximum batch size K	batch size, waiting time, departure time	exact	Özden et al. 2010 (under review)

Figure 3.2.: Elements for the analysis of batch building processes in material flow systems

gle pieces. It is assumed that the batch size of the arriving batch is described by an i.i.d. random variable. The departure process of the discrete time G|G|1-queue with batch arrivals is analyzed by Schleyer (2007). In addition, the distribution of the number of orders at the arrival instant can be determined, which can be used for the dimensioning of material flow buffers (see figure 3.3). Schleyer (2007) also presents mathematical models for the analysis of batch service queues. In production, batches of specific articles are produced, to reduce the number of necessary machine set ups. In some process steps, a certain number of units is treated in parallel, e.g in chemical washing, which equals a batch service. Schleyer (2007) investigates two different batch service strategies, the full batch policy, where orders arrive in batches and always a batch of a constant size K, which equals the maximum server capacity, is collected and served ($G^x|G^{[K,K]}|1$-queue), and the minimum batch size policy, where orders arrive in single units and a service process is initiated when at least L orders are accumulated in the queue ($G|G^{[L,K]}|1$-queue, also see Schleyer and Furmans 2007b). These elements are shown in figure 3.3. In order to analyze performance measures of the $G^x|G^{[K,K]}|1$-queue, Schleyer decomposed the system into two sub-

Element	Input	Output	Ex./ap.	author
G^X|G|1-queue with batch arrivals 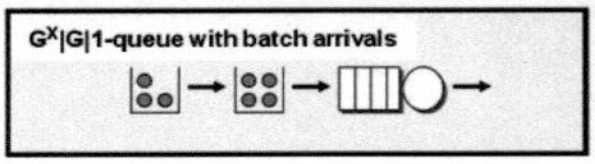	arrival time, batch size, service time	waiting time, nb. customer arrival, departure time	exact	Schleyer 2007, Schleyer and Furmans 2007a
G^X|G$^{[K,K]}$|1-queue 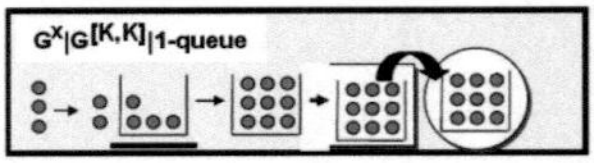	arrival time, batch size, service time, capacity K	waiting time, departure time	exact; decomposition: 2 subsystems	Schleyer 2007
G|G$^{[L,K]}$|1-queue 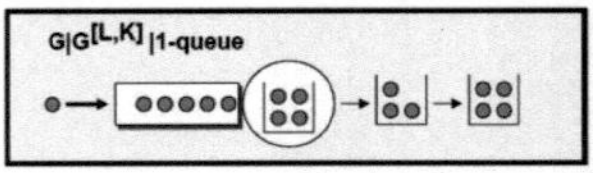	arrival time, service time	nb customer dep., waiting time, departure time, batch size	approximation, very close to simulation	Schleyer 2007, Schleyer and Furmans 2007b
G^X|G$^{[L,K]}$|1-queue 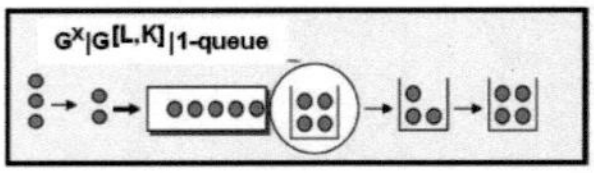	arrival time, batch size	nb customer dep., waiting time, departure time, batch size	exact	Özden and Furmans 2010
Sorting of batches 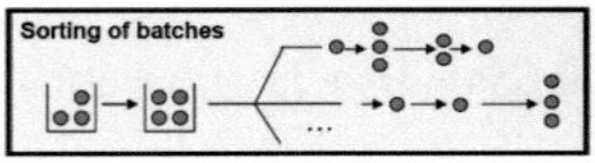	batch size, arrival time, relation of directions	departure time, batch size	exact	Schleyer 2007

Figure 3.3.: Elements for the analysis of batch queues and split of batches in material flow systems

systems, namely a collecting station running under the capacity rule and a G|G|1-queueing system. For the G|G$^{[L,K]}$|1-queue he presents an approximation method for the determination of the number of customers in the queue at the departure instant. The approximative results are very close to the results obtained by simulation. Given the number of customers in the departure instant, the interdeparture and waiting time distribution can be derived. The G|G$^{[L,K]}$|1-queue can be optimized choosing an optimal L depending on system costs such as operation and inventory costs. The model of Schleyer was extended by Özden and Furmans (2010) to a G^x|G$^{[L,K]}$|1-queue where the customers/orders arrive in batches. The authors present approximation methods, that are very close to the exact values, for the computation of the distribution of the number of customers in the departure instant, the waiting time and the interdeparture time distribution as well as the distribution of the departing batch size. Another operation in a manufacturing system is the sorting of batches that branches a batch arrival stream in several directions for a further processing, e.g. at the quality control. Given the stochastic description of the batch arrival stream and the branching probability, the stochastic streams after sorting can be calculated (see Schleyer 2007).

For each of the different batch queueing models shown in figure 3.3, mathematical approaches for the determination of the interdeparture time distribution exist, which enables us to investigate open queueing networks where batch processes are involved. Thus, a network can be composed from a given library of stochastic elements. This network can be analyzed and evaluated under various parameter configurations. The sojourn time distribution can be calculated for each node and a whole network. This gives us the probability that an order can be fulfilled in an acceptable time which is very crucial for the design of material flow and service systems. The mean number of material units/orders in the system can be estimated using Little's Law (Little 1961).

The presented models are well suited to analyze material flow systems. However, discrete time analysis also offers other reasonable fields of application. Inventory management, e.g. also benefits from methods dealing with empirical discrete probability distributions. Several works are modeling different inventory policies in discrete time scale to calculate the waiting time distribution of a customer order (see for example Zillus 2003, Tempelmeier 2006b, Tempelmeier 2006a and Tempelmeier and Fischer 2009). It is thus possible to set the parameters of an inventory policy in a way, that ensures a certain service level.

We have seen, that a lot of important discrete time queueing models are developed so far but there is one central element missing: the discrete time multi-server queueing system. Because of its numerous application possibilities, the multi-server queue in continuous time domain is well researched, as we will see in the following section.

3.2. Multi-Server Queueing Systems with General Arrival and Service Processes

Parameters of multi-server queueing systems are often calculated via system interpolations, where known parameters of basic systems - like $M|M|m$- or $M|D|m$-systems - are used to approximate values for more complex systems, e.g. $M|G|m$-systems. This approximations are often weighted averages. Lee and Longton (1959) develop a simple approximation for the mean waiting time in a $M|G|m$-system, based on a $M|M|m$-system. Björklund and Elldin (1964), Boxma et al. (1979) as well as Kimura (1986) and (1994) extend this approximation using two bases $(M|M|m, M|D|m)$.

In addition, Kimura (1994) generalizes his approximations for the $GI|G|m$-system. Compared to related approaches of Sakasegawa (1977) and Page (1972), Kimura (1994) obtains results, that are quite good for high utilizations and for interarrival and service time processes with low variabilities. However, the results get less accurate for increasing variabilities. Thus, Seelen and Tijms (1984) combine methods from Kleinrock (1976) for high variabilities and Burman and Smith (1983) for low variabilities to get good approximations for the mean waiting time in the whole range of variabilities.

Kingman (1970) and Brumelle (1971) develop an upper bound for the number of waiting customers in a $GI|G|m$-system and an upper bound for the mean waiting time. The methods are based on the upper bound of Kingman (1962) for single-server systems.

There also exist some approaches for the approximation of probability distributions for the waiting time or the number of customers in $GI|G|m$-queueing systems. Because of the high complexity, the generally distributed arrival and service processes are normally represented by phase type distributions or modifications which have the memoryless property. It is Neuts (1981) who first presents his matrix-analytic method to calculate the system state probabilities for a $PH|PH|1$-system. Breuer (2003) uses the matrix-analytic method to calculate the system state probabilities of a multi-server queue with phase type distributed processes.

Bertsimas (1988) presents a numerical method for the determination of the waiting time distribution for the $MGE|MGE|s$-system, where MGE is the class of mixed generalized Erlang probability density functions (pdfs), which is a subset of Coxian pdfs that have rational Laplace transform. The approach can get very complex, subject to the number of phases for arrival and service processes, but the obtained distributions are highly accurate. Choi et al. (2005) present a method for the calculation of the waiting time distribution in a multi-server system with a limited waiting room. They do not approximate the arrival and service processes by any known distribution, but make rough estimations for some parameters, which leads to imprecise approximations for the waiting time distribution, especially for low system utilizations. Kim et al. (2004) present an exact method to derive stationary queue length distributions for multi-server systems, where the arrival process is either Bernoulli or Poisson distributed and the service time is deterministic. They show that the stationary number of customers in a queueing system is the sum of two independent random variables, one of which is the stationary number of customers in the queue and the other is the number of customers that arrive during the time a customer spends in service.

For the determination of the queue length in a general multi-server system, Halfin and Whitt (1981) analyze $GI|M|m$-systems. They present approximations for the queue length, which tends to be best for queueing systems with high utilizations and a high number of parallel servers. Puhalskii and Reiman (2000) extend the approximation method to cases with phase type distributed service processes ($GI|PH|m$-systems). Whitt (2005) in turn extends the approach to multi-server systems with hypo-exponential distributed service processes, which are a combination of an exponential distribution and a point mass at zero. Each of these approaches is limited to memoryless service processes.

We have seen, that a lot of approaches for the performance analysis of multi-server queueing systems exist so far. But the methods either approximate input parameters by known distributions and get exact results for the output parameters, or they are based on general distributions and only mean values of the performance figures can be calculated.

We can summarize that there is no calculation method existing, that is adequate for the analysis of parallel servers in material flow systems. Thus, in the following chapters, we will develop analytical methods to derive the distributions of the performance measures presented in section 2.3.

4. Distribution of the Number of Customers at the Arrival Instant

> The real voyage of discovery consists not in making new landscapes but in having new eyes.
>
> *Marcel Proust*

The aim of this chapter is to calculate the distribution of the number of customers in a multi-server queueing system at the arrival instant of a customer. In the moment when a customer arrives, he finds a certain number of customers already present in the queueing system. If we have a limited buffer size, the arrival instant is the critical moment that can decide whether the arriving customer can be absorbed by the waiting room or has to be rejected. Knowing the distribution of the number of customers in the arrival instant can help us, for example, to size a buffer in front of a set of machines, in order to meet the stream of arriving customers/jobs.

We calculate the probabilities to see a certain system state at an arrival instant in section 4.1, using a homogenous discrete Markov chain that we embed at the arrival instants. Here, we can mention that many stochastic problems in material flow systems can be solved using a discrete homogenous Markov chain. One application can be found in Lippolt (2003), who calculates the expected travel times in automated storage/retrieval systems with double-deep storage. Schleyer (2007) uses discrete homogenous Markov chains to analyze batch processes in material flow systems. Another application of the approach is given by Matzka et al. (2009), who calculate the optimal number of kanbans in a heijunka leveled production systems, where the customers demand arrives at discrete time instants via a milk-run, and is regulated by a kanban loop, too.

To determine the system state probabilities, we first have to calculate the transition probabilities from a state at the arrival of a certain customer to the system state at the arrival of the succeeding customer in section 4.2. Therefore we will distinguish between three different cases according to the utilization of the system in the arrival moments.

4.1. Steady State Probabilities

In the moment when a customer arrives at the queueing system, he finds a certain number of customers present in the system. These customers are waiting or in process. As there are m servers in the system, a maximum number of m customers can be in process at the arrival instant, each with a certain residual service time. The system state π at the arrival moment of a customer can be defined as a $(m+1)$-tuple

$$\pi = (\eta, r_1, ..., r_m) = (\eta, \vec{r}) \tag{4.1}$$

$$with \quad \eta \in \mathbb{N}_0 \quad and \quad r_s \in \{0, 1, ..., \beta_{max} - 1\} \quad \forall s \in \{1, ..., m\}$$

where η is the number of customers present in the system at an arrival moment and $\vec{r} = (r_1, ..., r_m)$ are the residual service times of the m servers. We observe the system states at the moment immediately before the arrival instants. Thus, the number of customers at the arrival moment contains all customers waiting or in process, including customers that have finished their service in the arrival moment, excluding the customer that just arrives.

Figure 4.1 gives an example for the system state π at the arrival of a customer. At his arrival the customer finds $\eta = 3$ customers already present in the system. As there are $m = 2$ servers working in parallel, each of them has a customer in service with a residual service time

$(r_1 = 1, r_2 = 3)$. One customer is waiting in the waiting room. Thus, our system state π at the arrival of this customer is defined by the tuple $(\eta, r_1, r_2) = (3, 1, 3)$.

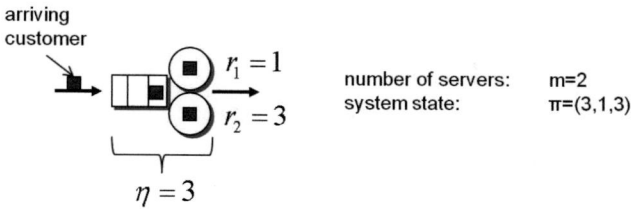

Figure 4.1.: exemplary system state π at the arrival of a customer

In the steady state, the system state π is identically distributed at each arrival instant. The state can be described by a random variable X, and the distribution is denoted by:

$$P(X = \pi) = x_\pi \tag{4.2}$$

A transition from one state to another state only takes place at discrete points of time, the arrival moments. We identify a Markov process and calculate the steady state probabilities x_π using a discrete homogeneous Markov chain that is embedded at the arrival instants. The state space of this Markov chain can be either finite or infinite, depending on the configuration of the system.

Let us consider the arrival moment of a certain customer c. Customer c sees a certain system state $\pi^c = i$ with $i = (\eta^c, r_1^c, ..., r_m^c)$ at his arrival. We also define a system state $\pi^{c+1} = j$ with $j = (\eta^{c+1}, r_1^{c+1}, ..., r_m^{c+1})$ that the succeeding customer $c + 1$ sees at his arrival instant. We introduce p_{ij} as the transition probability from state i at the arrival of a customer c to state j at the arrival of customer $c + 1$.

A transition from state i to state j and thus, the transition probabilities p_{ij} depend on the state i and the number of customers that can be served within the interarrival time interval α (see figure 4.2).

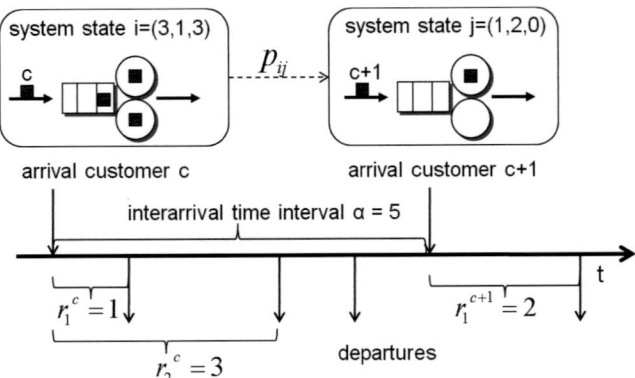

Figure 4.2.: transition from a state i to a state j

In our example, we assume that the customer in the queue will have a service time of $\beta = 6$ and customer c has a service time of $\beta = 1$. Thus, the queueing system serves three customers within the interarrival time interval $\alpha = 5$. Therefore, at the end of time interval α, server $s = 1$ is busy with $r_1^{c+1} = 2$, and the other server $s = 2$ is idle with $r_2^{c+1} = 0$.

The transition probabilities p_{ij} can be calculated as shown in the following sections. Knowing the transition probabilities p_{ij}, we will be able to calculate the state probabilities as follows:

$$x_j = \sum_i x_i \cdot p_{ij} \tag{4.3}$$

with

$$\sum_j x_j = 1 \tag{4.4}$$

Thus, we get an over-determined set of linear equations and we can calculate the state probabilities x_j using the transition probabilities p_{ij}. Since the summation index in equation 4.3 goes to infinity, because theoretically an infinite number of customers can be present in the system, we have to truncate the summation after an appropriate number of steps. Therefore, the method is exact within an ϵ-environment.

As in the steady state the state probabilities are identically distributed at each arrival moment, the state distribution at the arrival of an arbitrary customer x_π is identical to the distribution x_j of the system state at the arrival of customer $c + 1$.

So, the first step will be the determination of the transition probabilities p_{ij}.

4.2. Transition Probabilities

Let us consider a transition matrix P with its elements p_{ij} that give the probability for a transition from state i at the arrival of customer c to state j at the arrival of customer $c + 1$ (see figure 4.3). For a higher clarity, in figure 4.3 the system state is aggregated to the number of customers in the system (η^c and η^{c+1}). We can see the system state in detail at the left side of the matrix. The possible transitions are colored in the matrix and can be motivated as follows. From the arrival of customer c to the arrival of customer $c + 1$, only customer c can increase the number of customers in the system. Thus, if no service ends within the interarrival time interval, the number of customers in the system at the arrival of customer $c + 1$ can be at maximum $\eta_{max}^{c+1} = \eta^c + 1$. If some customers' services end within the interarrival time interval α, η^{c+1} will be smaller than this maximum.

As mentioned in section 4.1, the transition probabilities p_{ij} depend on the number of customers the system can serve within the interarrival time interval α. Thus, we introduce a random variable $U^{\alpha,\vec{r}^c,\vec{r}^{c+1}}$ for denoting the number of customers, the system can serve within a certain interarrival time interval α, starting from a system state with a certain combination of residual service times $\vec{r}^c$ and ending up in a combination $\vec{r}^{c+1}$. The probability vector of this variable is denoted by $\vec{u}^{\alpha,\vec{r}^c,\vec{r}^{c+1}}$ with the probabilities $u_v^{\alpha,\vec{r}^c,\vec{r}^{c+1}}$ to serve exactly v customers ($v = 0, ..., v_{max}$). As this distribution is only valid for a certain interarrival time interval α, we have to determine distribution $u_v^{\alpha,\vec{r}^c,\vec{r}^{c+1}}$ for

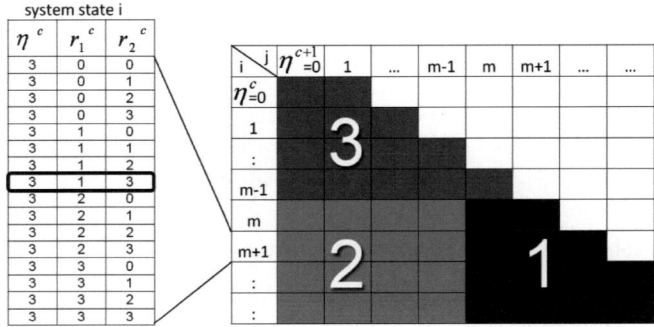

Figure 4.3.: transition matrix P with the transition probabilities p_{ij}

each possible value of α ($\alpha = 1, ..., \alpha_{max}$). When we know the distributions $u_v^{\alpha, \vec{r}^c, \vec{r}^{c+1}}$, the transition probabilities p_{ij} can be determined as follows:

$$p_{ij} = \sum_{\alpha=1}^{\alpha_{max}} a_\alpha \cdot u_{v=\eta^c+1-\eta^{c+1}}^{\alpha, \vec{r}^c, \vec{r}^{c+1}}, \tag{4.5}$$

where a_α is the probability that the time interval between the arrivals of customers c and $c + 1$ has a length of α. When a customer c sees a state i with η^c customers in the system at his arrival, and the next arriving customer $c+1$ sees a state j with η^{c+1} customers in the system, the system must have served $\eta^c + 1 - \eta^{c+1}$ customers in the time interval between the two arrivals. We have $\eta^c + 1$ because customer c is then counted as present in the system. The probabilities p_{ij} for a transition from a present state i at the arrival of a customer c to a state j at the arrival of customer $c + 1$ are depending on the time interval α between the two arrivals with its probability a_α and the probability that exactly $v = \eta^c + 1 - \eta^{c+1}$ customers can be served within this time interval α, starting from a certain system state i ending up in system state j.

The aim of the following sections is to calculate the probability distributions $u_v^{\alpha, \vec{r}^c, \vec{r}^{c+1}}$, which give the probabilities that the queueing system serves exactly v customers within the interarrival time interval α. As the system is not always fully occupied at each arrival moment, we have to distinguish between three cases (see the matrix in figure 4.3). In case 1, all servers are busy at the arrival of customer c ($\eta^c \geq m$) and they are still busy at the arrival of customer $c+1$ ($\eta^{c+1} \geq m$). The according calculation method for $u_v^{\alpha,\bar{r}^c,\bar{r}^{c+1}}$ can be found in section 4.2.1. In case 2 (section 4.2.2), all servers are busy at the arrival of customer c and at least one server is idle at the arrival of customer $c+1$ (see figure 4.2). In case 3, the arriving customer finds the system with $\eta^c < m$ customers in it, so some servers are idle. Thus, he does not have to wait, and can be processed immediately (see section 4.2.3).

4.2.1. Case 1: Transition from Busy System to Busy System

In the first case, all stations are busy at the arrival of customer c and they are still busy at the arrival of customer $c+1$ (an example is given in figure 4.4).

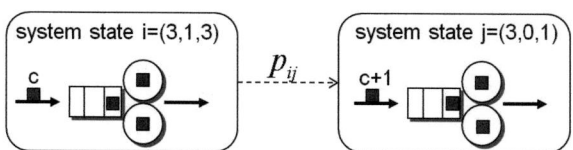

Figure 4.4.: transition from a state with $\eta^c \geq m$ to a state with $\eta^{c+1} \geq m$

The probability $u_{\eta^c+1-\eta^{c+1}}^{\alpha,\bar{r}^c,\bar{r}^{c+1}}$ that the system is able to serve $\eta^c + 1 - \eta^{c+1}$ customers within an interarrival time interval α, starting from a combination of residual service times $\bar{r}^c$ and ending up in a combination of residual service times $\bar{r}^{c+1}$, is the probability that the sum of customers served at all of the m servers is exactly $\eta^c + 1 - \eta^{c+1}$. The performance of the whole queueing system is the sum of the number of customers each server can serve. Thus, the probability distribution of the number of customers the whole system can serve, is a convolution of the probability distributions for each of the single servers with their special combination of residual service times:

$$\overrightarrow{u}^{\alpha,\vec{r}^c,\vec{r}^{c+1}} = \overrightarrow{u}^{\alpha,r_1^c,r_1^{c+1}} \otimes \overrightarrow{u}^{\alpha,r_2^c,r_2^{c+1}} \otimes \ldots \otimes \overrightarrow{u}^{\alpha,r_m^c,r_m^{c+1}} \tag{4.6}$$

To determine the performance of the whole system, we first have to know how many customers a single station can serve within a certain interarrival time interval.

Performance of a single server

If a server is busy at the arrival moment of customer c, one customer is already in service and has a residual service time r^c. The customer in process can only have left the system at the arrival of the next customer, if the residual service time is smaller than the interarrival time. The server is then able to serve at least one customer in the given interval α. If the residual service time is greater than or equal to the interarrival time, the server is not able to finish the service of the current customer and he is still in process at the arrival of the succeeding customer. Thus, the probabilities that one server is able to serve a certain number of customers can be calculated according to these 2 different cases:

if $(r^c \geq \alpha)$:

$$u_v^{\alpha,r^c,r^{c+1}} = \begin{cases} 1 & \text{if } v = 0 \wedge r^{c+1} = r^c - \alpha \\ 0 & else \end{cases}$$

In the case of a residual service time greater than the interarrival time interval α, the customer with residual service time r^c is still in the system at the arrival of the next customer and has then a residual service time of r^{c+1} with $r^{c+1} = r^c - \alpha$ (see figure 4.5).

if $(r^c < \alpha)$:

$$u_v^{\alpha,r^c,r^{c+1}} = \begin{cases} 0 & \text{if } v = 0 \\ b_{r^{c+1}+\alpha-r^c} & \text{if } v = 1 \\ \sum_{\Delta=1}^{\phi}(b_{r^{c+1}+\Delta} \cdot b_{\alpha-\Delta-r^c}^{(v-1)\otimes}) & \text{if } 2 \leq v \leq v_{max} \end{cases}$$

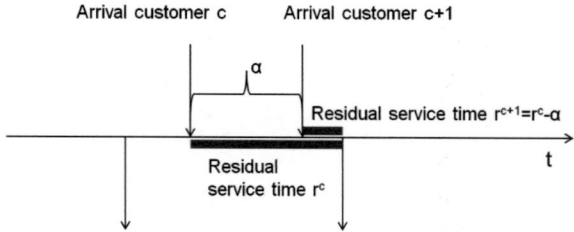

Figure 4.5.: residual service time is greater than or equal to the interarrival time interval

$$with \quad v_{max} = \lfloor (\alpha - \Delta - r^c)/\beta_{min} \rfloor + 1$$

$$and \quad \phi = min\{\alpha - r^c - \beta_{min}; \beta_{max} - r^{c+1}\}$$

In the case that the residual service time is smaller than the interarrival time, a new job will start, before the next customer arrives (see figure 4.6). We look at all possible cases to reach a residual service time

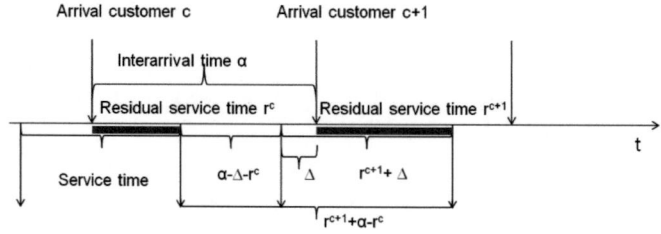

Figure 4.6.: residual service time is smaller than the interarrival time interval

r^{c+1}. If only one customer starts his job, his service time β has to be $r^{c+1} + \alpha - r^c$ to get a residual service time of r^{c+1} time instants. The corresponding probability for this service time is $b_{r^{c+1}+\alpha-r^c}$. If more than one customer started his job during the interarrival time interval α, the last customer has to have a service time greater than the residual service time r^{c+1}. The difference between this service time and the residual service time is denoted by Δ and has to be at least 1 time increment. Furthermore, Δ should be at most $\alpha - r^c - \beta_{min}$ to leave time for another customer's service and Δ can be at most $\beta_{max} - r^{c+1}$.

4.2.2. Case 2: Transition from Busy System to (Partly) Idle System

In the second case, all servers are busy at the arrival of customer c and at least one server is empty at the arrival of customer $c+1$ (an example is given in figure 4.7). This implies, that each of the empty servers has been able to serve at least one customer and then had an idle time.

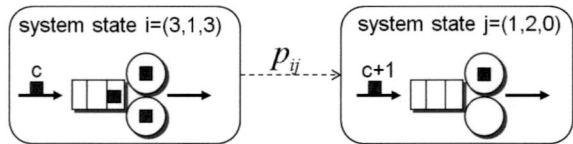

Figure 4.7.: transition from a state with $\eta^c \geq m$ to a state with $\eta^{c+1} < m$

Because of the unknown value of the idle time, we can not use the same method as for the first case ($\eta^c \geq m \wedge \eta^{c+1} \geq m$). We thus have to determine the transition probabilities by allocating each customer to one of the m servers according to the servers availability and the customers' service time.

For each server s, we introduce a working time account (wta) which counts the amount of working time, the server needs to serve his allocated customers. We initialize each working time account with the residual service time of the customer currently in process:

$$wta_s = r^c_s \quad \forall s \in \{1, ..., m\} \tag{4.7}$$

A total of m customers is already in process. The customers in the queue ($\eta^c - m$) and the customer that just arrived, have to wait. We introduce a variable q, with $q = 1, ..., \eta^c - m + 1$, to label the customers that are present in the queue. Each customer in the queue has a certain service time β^q. We imagine a set of possible service time combinations $\vec{t} = (\beta^{q=1}, ..., \beta^{q=\eta^c - m + 1})$ for the $\eta^c - m + 1$ waiting customers. Each customer has a range for his service time β between 1 and β_{max} time units. So for $\eta^c - m + 1$ waiting customers we get $(\beta_{max})^{\eta^c - m + 1}$ possible combinations of service times. For each combination $\vec{t}$, we can assign the customers to the m servers by following the rule that the first customer in the queue is allocated to the first available server. The minimum of the wta then increases by the service time β^q of the currently regarded customer. We identify 5 steps to calculate the wta.

1. set $q := 0$

2. determine server z such that $wta_z = \min_s \{wta_s\}$

3. set $wta_z := wta_z + \beta^q$

4. set $q := q + 1$

5. if $q \leq \eta^c - m + 1$ continue with step 2, else stop.

Following these steps until all waiting customers are assigned to one of the m servers, we get the final wta for each server.

Figure 4.8 gives an example of the determination of working time accounts. We have 2 servers in the exemplary queueing system. Both are busy at the arrival of customer c. We initialize each wta with

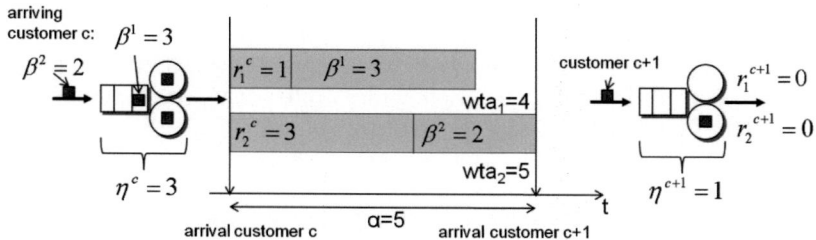

Figure 4.8.: example for the determination of working time accounts

the corresponding residual service time we know from our initial state $i = (3, 1, 3)$. We have to assign two waiting customers to the servers and do that for the set of service times $\vec{t} = (3, 2)$. As wta_1 is smaller than

wta_2, we assign customer $q = 1$ to server 1 and increase wta_1 by his service time $\beta = 3$. We assign customer $q = 2$ to server 2, accordingly.

Comparing the wta to the given interarrival time interval, we can find out which server is busy and which is idle at the arrival of customer $c+1$. Servers with $wta < \alpha$ are idle at the arrival of customer $c+1$, and have a residual service time of $r^{c+1} = 0$. If $wta \geq \alpha$, the server is busy and has a residual service time of $r^{c+1} = wta - \alpha$. As the interarrival time interval of our example has a length of $\alpha = 5$ time units, we obtain that server 1 is idle at the arrival of customer $c+1$ ($r_1^{c+1} = 0$) and server 2 is busy with a residual service time $r_2^{c+1} = 0$.

The number of remaining customers η^{c+1} at the arrival of customer $c+1$ is equal to the number of servers with $wta \geq \alpha$. So we know the number of customers served within the interarrival time interval α and we know the residual service time of each server. We thus can determine the system state j at the arrival of customer $c+1$. Knowing the probabilities for the service time of each customer and the probability of the given inter arrival time interval α, the probability for a transition from state i to state j can be calculated as follows:

$$p_{ij} = \prod_{q=1}^{\eta^c - m + 1} b_{\beta q} \cdot a_\alpha \tag{4.8}$$

Performing these steps for each possible combination of $\vec{t}$ and each possible interarrival time interval α, we can find out every possible transition and its probability p_{ij} in the range of $(\eta^c \geq m \wedge \eta^{c+1} < m)$.

4.2.3. Case 3: Transition from (Partly) Idle System to (Partly) Idle or Busy System

In the third case, the arriving customer c finds the system with $\eta^c < m$ customers in it. Thus, he does not have to wait, and can be processed immediately. Figure 4.9 illustrates an example for this case. Servers s=1 and s=2 are idle at the arrival of customer c and server s=3 is busy. One of the idle servers starts the process of customer c immediately after his arrival (server s=1 in the example), the other servers stays idle (server s=2).

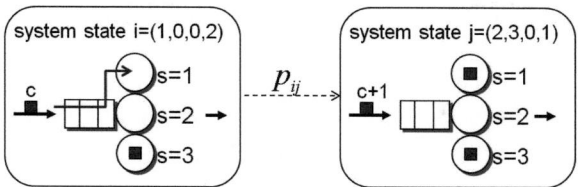

Figure 4.9.: transition from a state with $\eta^c < m$ to a state with $\eta^{c+1} \leq m$

The distribution of the number of customers one single server can serve, can thus be calculated according to these 3 different cases:

I: Server is already busy at arrival

When a server is already busy at the arrival of the new customer (as server s=3 in figure 4.9), he can finish the service, if the residual service time is smaller than the interarrival time α. If the residual service time is equal to or greater than the interarrival time, the customers stays in process and it remains a residual service time of r^{c+1}. In this case, zero customers have been served.

if $(r^c \geq \alpha)$:

$$u_{I,v}^{\alpha,r^c,r^{c+1}} = \begin{cases} 1 & \text{if } v = 0 \ \wedge \ r^{c+1} = r^c - \alpha \\ 0 & else \end{cases}$$

If the residual service time is smaller than the interarrival time, the server finishes the service of one customer ($v = 1$) and the residual service time is zero.

if $(r^c < \alpha)$:

$$u_{I,v}^{\alpha,r^c,r^{c+1})} = \begin{cases} 1 & \text{if } v = 1 \ \wedge \ r^{c+1} = 0 \\ 0 & else \end{cases}$$

II: Server is idle at arrival and starts service of arriving customer

The customer that just arrives, enters an idle server of the system (server s=1 in figure 4.9) and starts his process immediately. Customer c is finished when the next customer arrives, if his service time β (with its probability of appearance b_β) is smaller than the interarrival time α, else he is still in process. The probability distribution of the number of customers this initially empty station can serve within an interarrival time interval α can be determined as follows:

if $(\beta \geq \alpha)$:

$$u_{II,v}^{\alpha,r^c,r^{c+1}} = \begin{cases} b_\beta & \text{if } v = 0 \wedge \beta = r^{c+1} + \alpha \wedge 0 \leq \beta \leq \beta_{max} \\ 0 & else \end{cases}$$

if $(\beta < \alpha)$:

$$u_{II,v}^{\alpha,r^c,r^{c+1}} = \begin{cases} \sum_{\beta=0}^{\alpha-1} b_\beta & \text{if } v = 1 \wedge r^{c+1} = 0 \\ 0 & else \end{cases}$$

III: Server is idle at arrival and stays idle

The servers that are already idle at the arrival of customer c and stay idle until the arrival of the next customer, can not serve any customer (see server s=2 in figure 4.9). They start from and end up in a residual service time of zero.

$$u_{III,v}^{\alpha,r^c,r^{c+1}} = \begin{cases} 1 & \text{if } v = 0 \wedge r^{c+1} = 0 \\ 0 & else \end{cases}$$

Performance of the whole system

Knowing the probability distributions of the three possible cases (I, II, III), we can now determine the distribution for the performance of the whole queueing system. We first have to know, if an initial state is valid or not. An initial state $i = (\eta^c, r_1^c, ..., r_m^c)$ is valid, if the number of servers with a residual service time $r_s^c > 0$ is smaller than or equal to the number of customers present in the system (η^c). If this is not the case, all transition probabilities starting from this state i are equal to zero.

For a valid initial state, the distribution of the performance of the whole queueing system can be determined by a convolution of the distributions of the number of customers each single server can serve, according to the 3 different cases. So, if we start from a state with η^c customers in process, we have to convolute η^c vectors of case I with one vector of case II and $m - \eta^c - 1$ vectors of case III. In our example in figure 4.9 we have $\eta^c = 1$ customer in a system with $m = 3$ servers. We thus would convolute one vector of case II (for the idle server s=1 that serves customer c) with one vector of case III (for server s=2 that stays idle) and one vector of case I (for the busy server s=3).

4.3. Distribution of the Number of Customers at the Arrival Instant

Knowing the transition probabilities p_{ij}, we can now calculate the system state probabilities x_j and accordingly x_π using equations 4.3 and 4.4. The distribution of the system state is useful for the calculation of the waiting time distribution in the next chapter, but to get an information about the number of customers in the arrival instant, we first have to aggregate the system state probabilities x_π. Therefore, we introduce a random variable N for the number of customers in the system at the arrival instant. The probabilities to find exactly η customers in the system are denoted by $P(N = \eta) = n_\eta$. Summing up all state probabilities x_π for states with the same number of customers in the system, we get the distribution of the number of customers at the arrival instant n_η:

$$n_\eta = \sum_{r_1=0}^{\beta_{max}-1} \cdots \sum_{r_m=0}^{\beta_{max}-1} x_{(\eta,r_1,\ldots,r_m)} \quad \forall \eta \in \mathbb{N}_0 \tag{4.9}$$

According to this calculation, arbitrarily high values for the number of customers in the system can be possible. To constrain the distribution, we define a maximum value η_{max} with

$$\eta_{max} < \epsilon \qquad (4.10)$$

We choose ϵ according to the desired accuracy of calculation. In order to determine η_{max}, we iteratively increase or decrease this value starting from an arbitrary chosen initial value and calculate the system state probabilities x_π, until equation 4.10 is fulfilled.

5. Computation of the Waiting Time Distribution

> All things come to those who wait.
>
> *Marie Curie*

In the previous chapter we determined the distribution of the number of customers in the system at the arrival instant of a certain customer respectively the system state probabilities. Based on the distribution of the system state we will now determine the distribution of the waiting time, a customer will spend in the queue in front of the parallel servers. Knowing the waiting time distribution we will then be able to calculate the distribution of the sojourn time of a customer in the system. As mentioned before, the waiting time distribution and especially the sojourn time distribution are necessary to make conclusions about an on-time order fulfillment we can guarantee a customer. We also need the sojourn time distribution to know the replenishment time of material in a supermarket, for example.

5.1. Waiting Time Distribution

In the moment, when a customer c arrives at the queueing system, he either gets in process immediately or he has to wait in a queue. To describe the waiting process, we introduce the random variable W with its probability distribution $w_\omega, \omega = 0, ..., \infty$.

If an arriving customer c finds the system with $\eta^c < m$ customers in it, at least one server is idle, and he can get in process immediately. Then his waiting time ω is equal to zero. The probability for a waiting time of zero time units can thus be initialized as follows:

$$w_\omega = \sum_{\eta=0}^{m-1} \sum_{r_1=0}^{\beta_{max}-1} \cdots \sum_{r_m=0}^{\beta_{max}-1} x_{(\eta, r_1, \ldots, r_m)} \tag{5.1}$$

The waiting time of customer c can also be equal to zero, if at least $\eta^c - m + 1$ servers have a residual service time $r_s = 0$ and finish a service at the arrival moment of customer c. The probability for a waiting of $\omega = 0$ is thus higher than calculated with equation 5.1. We increase the probability for a waiting time of zero time units to the exact value, using the method we introduce in the following paragraph.

When customer c arrives in a system where all servers are currently occupied ($\eta^c \geq m$), he has a waiting time of $\omega \geq 0$. As mentioned before, the waiting time is zero, if $\eta^c = m$ and at least one server finishes his service at the arrival instant. To determine the waiting time distribution we use a method similar to the method introduced in section 4.2.2. For the system states i with $\eta^c \geq m$ we allocate all customers η^c that are present in the system to the m servers. For each server s, $s = 1, \ldots, m$ we again use a working time account (wta) which counts the amount of working time, the server needs to serve his allocated customers. Like in section 4.2.2, we initialize each working time account with the residual service time of the customer currently in process (customers 1 to m):

$$wta_s := r_s^c \quad \forall s \in \{1, \ldots, m\} \tag{5.2}$$

To calculate the waiting time of an arriving customer c, we now only have to regard the predecessors of customer c, because they influence the waiting time of customer c. For the waiting predecessors $q = 1$ to $q = \eta^c - m$ we imagine a set of possible service time combinations $\vec{t} = (\beta^{q=1}, \ldots, \beta^{q=\eta^c-m})$ for these $\eta^c - m$ customers that are the predecessors of customer c, where β^q is the service time of a certain customer q in the queue. Each customer has a range for his service time β between 1 and β_{max} time units. So for $\eta^c - m$ waiting customers we get $(\beta_{max})^{\eta^c-m}$ possible combinations of service times. For each combination $\vec{t}$, we can assign the customers to the m servers, following the rule that the first customer in the queue is allocated to the first available server. The minimum of the wta then increases by the service time β^q of the currently regarded customer. We again use 5 steps to calculate the wta, this time only regarding the waiting predecessors of the arriving customer c.

1. set $q := 0$

2. determine server z such that $wta_z = \min_{s}\{wta_s\}$

3. set $wta_z := wta_z + \beta^q$

4. set $q := q + 1$

5. if $q \leq \eta^c - m$ continue with step 2, else stop.

Following this rule until all predecessors of customer c are assigned to one of the m servers, we get a final wta for each server. The waiting time of customer c is then equal to the minimum wta_s of all servers $s = 1, ..., m$:

$$\omega := \min_{s}\{wta_s\} \tag{5.3}$$

Figure 5.1 gives an example for the determination of the working time accounts and the according waiting time ω. We have 2 servers in the exemplary queueing system. Both are busy at the arrival of customer c. We initialize each wta with the corresponding residual service time we know from our initial state i. We have to assign two waiting customers to the servers and do that for a set of service times $\vec{t} = (1, 3)$. In the end, $wta_1 = 3$ is smaller than $wta_2 = 4$. So, customer c will start his process in server 1 after a waiting time of $\omega = 3$.

Doing this allocation for each possible system state i at the arrival of customer c and each possible combination of service times $\vec{t}$ for the predecessors of customer c, we can determine the waiting time distribution w_ω by each time increasing the probability for a certain waiting time ω as follows:

$$w_{\omega=wta_{min}} := w_{\omega=wta_{min}} + x_i \cdot \prod_{q=1}^{\eta^c - m} b_{\beta^q} \tag{5.4}$$

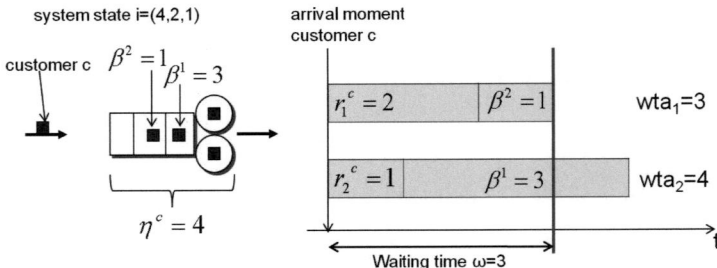

Figure 5.1.: Example for the determination of working time accounts and waiting time

5.2. Sojourn Time Distribution

Material flow systems should be designed in a way that it guarantees the order fulfillment in a predetermined time with a chosen probability (e.g. 95%). The distribution of the time an order remains in a system, defined as sojourn time, has to be known in order to determine its quantiles.

We introduce a random variable K for the sojourn time with its probability distribution $k_\kappa, \kappa = 0, ..., \infty$. The sojourn time of a customer in a queueing system is simply the sum of waiting time and service time. So, as we already calculated the distribution of the waiting time and we know the distribution of the service time, we can determine the sojourn time distribution by a convolution of this two known distributions:

$$\vec{k} = \vec{w} \otimes \vec{b} \tag{5.5}$$

6. Computation of the Interdeparture Time Distribution

> Nothing is particularly hard if
> you divide it into small jobs.
>
> *Henry Ford*

To be able to link the multi-server queueing element to other nodes of a network, we have to know the interdeparture time distribution. The interdeparture time distribution serves as interarrival time distribution for a succeeding node in a material flow network. The departure stream of a multi-server queue $\vec{d}$ can be seen as a merge of the departure streams of the m parallel servers $(\vec{d_1}, ..., \vec{d_m};$ see figure 6.1). As none of the servers

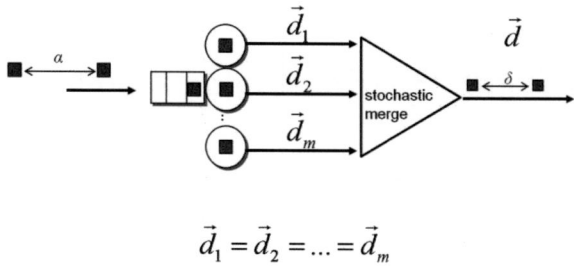

$$\vec{d_1} = \vec{d_2} = ... = \vec{d_m}$$

Figure 6.1.: Merge of output streams from a multi-server system

is prioritized by the customers, each stream of departing customers is identically distributed. Like in the single-server case, the interdeparture time intervals depend on the utilization of the system. If the system is

49

highly utilized and customers are waiting in the queue, the interdeparture time intervals of a single server are equal to the service time. If the utilization is lower and a server has idle time periods, the time interval between two departures increases by this idle time. In contrast to the single-server case, in a multi-server queue the idle time period of a server does not always end, when a new customer arrives in the system. If several servers are idle at this moment, the customer randomly chooses one of the idle servers. For this server the idle time period ends, for all others the idle time period continues at least for one interarrival time interval α. Taking this into account, we can calculate the idle time distribution of a single server in the multi-server system.

The computation of the interdeparture time distribution can be divided into 3 main steps. First, we have to calculate the idle time distribution of a single server as presented in section 6.1. Based on this distribution, we are able to calculate the interdeparture time distribution of a single server of the queueing system in section 6.2. Knowing the behavior of the single output streams, we can determine the interdeparture time distribution of the whole multi-server queueing system in section 6.3. As the interdeparture stream of the multi-server system is determined by an approximation method, we analyze the approximation quality of this method in section 6.4.

6.1. Idle Time Distribution of a Single Server

The idle time of a server starts, when at the departure instant of a served customer no successors are waiting in the queue. The server then stays idle at least until the next customer enters the queueing system (see figure 6.2).

The idle time period can either be finished, if the next customer starts his service in the regarded server, or it continues until the arrival of a further customer. Thus, we first have to calculate the distribution of the initial idle time, which is the time interval between the departure of the last customer from a single server and the arrival of the next customer in the multi-server system.

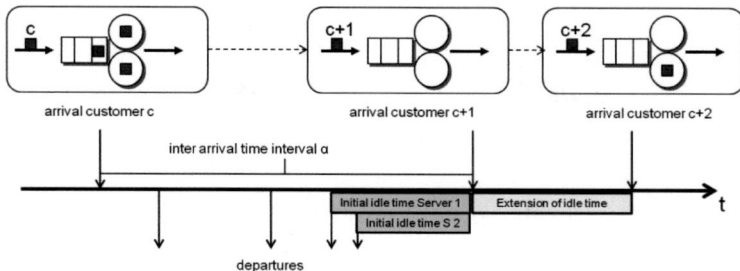

Figure 6.2.: Development of the idle time in a multi-server queueing system

Initial idle time distribution

The initial idle time is denoted by τ^I with $\tau^I \in \{1, ..., \alpha_{max}\}$, and its probability distribution is denoted by $y_{\tau^I}^I$. To calculate the initial idle time distribution, we first introduce an auxiliary distribution, denoted by $y_{\tau^{I*}}^{I*}$ with $\tau^{I*} \in \{0, ..., \alpha_{max}\}$. In this distribution we take into account every departure, whether it has a successor or not. So, there can also occur initial idle time values of zero, when a departing customer has a successor. This way, we can simultaneously calculate the initial idle time distribution and the probability for the appearance of an idle time after the departure of a certain customer $(1 - y_0^{I*})$ respectively the probability for no idle time after the departure (y_0^{I*}). These probabilities are used to calculate the interdeparture time distribution of a single server in section 6.2 (formula 6.14). To calculate $y_{\tau^{I*}}^{I*}$, we iterate over all possible system states i, all possible interarrival time intervals α and all possible service times of the customers in the system.

For each system state i with $\eta^c < m$, at least one server is empty and the arriving customer c can get in process immediately. He will leave the system within the interarrival interval α if his service time β is smaller than α. The server then is idle and the probability of the corresponding initial idle time can be increased as follows:

$$y_{\alpha-\beta}^{I*} := y_{\alpha-\beta}^{I*} + x_i \cdot a_\alpha \cdot b_\beta \tag{6.1}$$

For all servers that were busy at the arrival of customer c we have to distinguish between servers with a residual service time $r > 0$ and servers with residual service time $r = 0$. Servers with a residual service time that is greater than zero, will have an idle time period if the residual service time r is smaller than the interarrival time interval α. The probability of the corresponding initial idle time can be increased as follows:

$$y^{I*}_{\alpha-r} := y^{I*}_{\alpha-r} + x_i \cdot a_\alpha \cdot b_\beta \tag{6.2}$$

If a busy server has a residual service time of $r = 0$, it is able to start the service of the arriving customer as well as each of the empty servers. If the arriving customer chooses to start his service in one of the busy servers with residual service time $r = 0$, the idle time of one of these servers will equal zero. Then, the according initial idle time probability increases as follows:

$$y^{I*}_0 := y^{I*}_0 + x_i \cdot a_\alpha \cdot b_\beta \cdot \frac{\Xi}{m - \eta^c + \Xi} \cdot 1, \tag{6.3}$$

where Ξ is the number of busy servers with residual service time $r = 0$ and $(m - \eta^c)$ is the number of idle servers.

The initial idle time intervals of the other busy servers with residual service time $r = 0$ will be equal to the interarrival time interval α:

$$y^{I*}_\alpha := y^{I*}_\alpha + x_i \cdot a_\alpha \cdot b_\beta \cdot \frac{\Xi}{m - \eta^c + \Xi} \cdot (\Xi - 1) \tag{6.4}$$

If the customer decides to start his process in one of the $(m - \eta^c)$ empty servers, all busy servers with residual service time $r = 0$ will have an initial idle time period equal to the interarrival time interval α:

$$y^{I*}_\alpha := y^{I*}_\alpha + x_i \cdot a_\alpha \cdot b_\beta \cdot \frac{m - \eta^c}{m - \eta^c + \Xi} \cdot \Xi \tag{6.5}$$

For each system state i with $\eta^c \geq m$ we allocate all customers η^c that are present in the waiting room to the m servers. For each server s, we again introduce a working time account (wta). For each customer, that leaves the system within the interarrival time interval α, we can find out, if the customer has a successor from the waiting room ($\tau^{I*} = 0$), or if there is no successor and the server stays idle ($\tau^{I*} > 0$). If there is a successor, we increase the probability for an initial idle time of zero as follows:

$$y_0^{I*} := y_0^{I*} + x_i \cdot a_\alpha \cdot \prod_{q=1}^{\eta^c - m + 1} b_{\beta q} \tag{6.6}$$

If a departing customer has no successor, the initial idle time τ^{I*} is the difference between the interarrival time interval α and the working time account of the according server:

$$\tau^{I*} = \alpha - wta \tag{6.7}$$

For each departing customer without successor, we can increase the corresponding idle time probability as follows:

$$y_{\alpha-wta}^{I*} := y_{\alpha-wta}^{I*} + x_i \cdot a_\alpha \cdot \prod_{q=1}^{\eta^c - m + 1} b_{\beta q} \tag{6.8}$$

The idle time distribution $y_{\tau^I}^I$ under the condition, that the idle time is greater than zero, can now be determined as follows:

$$y_{\tau^I}^I = \begin{cases} 0 & \text{if } \tau^I = 0 \\ \dfrac{y_{\tau^{I*}}^{I*}}{1 - y_0^{I*}} & \text{if } \tau^I = 1, ..., \alpha_{max} \end{cases}$$

Probabilities for ending or continuing idle time period

Starting from an initial idle time τ^I, a server will have a busy period again, when the next arriving customer enters the regarded server. The probability for this transition is denoted by $p_{idle \to busy}$. All other servers, that are idle, will then stay idle with the complementary probability $p_{idle \to idle}$. We have to regard all system states i where the number of customers present in the system is less than the number of servers ($0 \leq \eta^c < m$) and thus, some servers are idle. The Ξ servers that are busy but with a residual service time of $r = 0$ are also able to start the service of the arriving customer c, so this customer can choose between $m - \eta^c + \Xi$ servers. If the arriving customer c joins one of the empty servers $(m - \eta^c)$, for this server the idle time ends, and for all other empty servers $(m - \eta^c - 1)$, the idle time period continues. If the arriving customer c joins one of the Ξ busy servers with residual service time of zero, all idle servers stay idle. Therefore the probabilities $p_{idle \rightarrow busy}$ and $p_{idle \rightarrow idle}$ can be calculated as follows:

$$p_{idle \rightarrow busy} = \frac{\sum_{\eta^c=0}^{m-1} n_{\eta^c} \cdot \frac{m-\eta^c}{m-\eta^c+\Xi} \cdot 1}{\sum_{\eta^c=0}^{m-1} n_{\eta^c} \cdot \left(\frac{(m-\eta^c)(m-\eta^c)}{m-\eta^c+\Xi} + \frac{\Xi(m-\eta^c)}{m-\eta^c+\Xi} \right)} \tag{6.9}$$

$$p_{idle \rightarrow idle} = \frac{\sum_{\eta^c=0}^{m-1} n_{\eta^c} \cdot \left(\frac{(m-\eta^c)(m-\eta^c-1)}{m-\eta^c+\Xi} + \frac{\Xi(m-\eta^c)}{m-\eta^c+\Xi} \right)}{\sum_{\eta^c=0}^{m-1} n_{\eta^c} \cdot \left(\frac{(m-\eta^c)(m-\eta^c)}{m-\eta^c+\Xi} + \frac{\Xi(m-\eta^c)}{m-\eta^c+\Xi} \right)} \tag{6.10}$$

Idle time distribution

Knowing the probabilities for an ending or continuing idle time period, we can calculate the idle time distribution vector $\vec{y}$, according to the method for the calculation of the interdeparture time distribution of a stochastic split (see Furmans 2004):

$$\begin{aligned}
\vec{y} = \; & p^0_{idle \rightarrow idle} \cdot p_{idle \rightarrow busy} \cdot \vec{y}^I \\
& + p^1_{idle \rightarrow idle} \cdot p_{idle \rightarrow busy} \cdot \vec{y}^I \otimes \vec{a} \\
& + p^2_{idle \rightarrow idle} \cdot p_{idle \rightarrow busy} \cdot \vec{y}^I \otimes \vec{a}^{2\otimes} \\
& \vdots
\end{aligned} \tag{6.11}$$

The idle time period of a certain server can either end with the arrival of the next customer, or it can go on for another interarrival time interval. The idle time distribution $\vec{y}$ is thus a weighted convolution of the initial idle time distribution $\vec{y}^I$ and the distribution of several succeeding interarrival time intervals.

According to this calculation, arbitrarily high values of idle time periods can be possible. To constrain the distribution of the idle time, we define a maximum value τ_{max} with

$$1 - \sum_{\tau=1}^{\tau_{max}} y_\tau < \epsilon \tag{6.12}$$

We choose ϵ according to the desired accuracy of calculation and attach the remaining probability mass to the last element of the distribution.

$$y_{\tau_{max}} = 1 - \sum_{\tau=1}^{\tau_{max}-1} y_\tau \tag{6.13}$$

This way we guarantee a minor influence to the mean value of the idle time compared to an even distribution of the probability mass to all possible values.

6.2. Interdeparture Time Distribution of a Single Server

The interdeparture time distribution of a single server can now be determined according to two possible cases. If a departing customer has a successor, the interdeparture time is equal to the service time, and thus, identical distributed. In the second case, the departing customer leaves the server empty, so the interdeparture time is the sum of idle time and service time of the next entering customer. The interdeparture time distribution of a single server can thus be calculated as follows:

$$\vec{d}_{single} = y_0^{I*} \cdot \vec{b} + (1 - y_0^{I*}) \cdot \vec{y} \otimes \vec{b} \tag{6.14}$$

6.3. Interdeparture Time Distribution of the Multi-Server System

The output streams of the m parallel servers are merged to one stream of customers that leave the queueing system. The departure process can thus be modeled as a stochastic merge of m arriving streams from the single servers (see figure 6.1). Knowing the interdeparture time distribution of a single server, the distribution of the multi-server system can easily be calculated using the approximation method for a stochastic merging element (see Furmans 2004). Furmans proposes to use a renewal process as an approximation of the resulting point process in the merged stream. This approximation is based on the idea that the residual interdeparture time in a merged stream is the minimum of the residual interarrival times of two arriving streams. From the distribution of the residual interdeparture time, the interdeparture times in the merged stream can be regained using their expected value. To calculate the interdeparture time distribution of a multi-server system, we first merge 2 streams according to the mentioned method. For m output streams, we iteratively merge the obtained stream with another single stream, until all m streams are included.

6.4. Analysis of the Approximation Quality

With the presented method we obtain exact results for the interdeparture time distribution of a single server (equation 6.14). In the next step, we use an approximation method for the calculation of the interdeparture time intervals in the merged stream and we assume to merge independent output streams of m servers which is not true for several examples. This approximation has no influence on the mean value of the interdeparture time distribution, but the variability can differ. Thus, we want to estimate the quality of the presented method comparing the interdeparture time variability c_d^2 from our analytical method to simulation results (see table 6.1). We choose a simulation length of 10 million served customers and estimate the relative error of the approximation results compared to simulation results. Note, that this simulation length causes relative errors of 0.1% comparing the waiting time variability obtained by simulation to the exact results from our calculation method.

For the estimation of the interdeparture time variability in $G|G|m$-systems, Whitt (1983) proposes the following approximation:

$$c_d^2 = 1 + (1 - \rho^2)(c_a^2 - 1) + \frac{\rho^2}{\sqrt{m}}(c_b^2 - 1) \qquad (6.15)$$

The formula shows that for systems with a low utilization ρ, the interdeparture time variability c_d^2 is highly influenced by the interarrival process, while for high utilizations, the service time variability c_b^2 is the contributing factor. We thus regard queueing systems with high and low utilizations to quantify the influence of interarrival and service time variability. Additionally, we compare results for two and three servers in order to identify the influence of the number of servers on the approximation quality (see table 6.1).

For a low utilization of $\rho = 0.5$ the interarrival time variability c_a^2 has a strong influence on the approximation quality. When c_b^2 is low (0.0 or 0.1), the departure instants from one server are highly dependant on the departures from the other servers and the streams are not independent. The merge thus causes high deviations of c_d^2 compared to the exact results from simulation experiments. For increasing c_b^2, the error decreases. The influence of the service time variability c_b^2 is secondary, but also visible in the experiments.

For a high utilization of 0.9 the approximation quality is mainly influenced by the service time variability c_b^2. When c_a^2 is low (0.0 or 0.1), the departure instants from one server are highly depending on the departures from the other servers and the streams are not independent. The merge thus causes high deviations of c_d^2 compared to the exact results from simulation experiments. For increasing c_a^2, the error decreases, but the influence of the service time variability c_b^2 is still visible in the experiments.

Increasing the number of servers from $m = 2$ to $m = 3$, the dependencies between the single streams increase, because the customer has more options to choose from. Therefore, the influence of low values of c_a^2 and the dependencies decrease. This can be seen in table 6.1 especially for $\rho = 0.5$, where the influence of the interarrival time variability on the output stream is stronger. Comparing the errors for $m = 2, \rho = 0.5, c_a^2 = 0$ to the results for $m = 3, \rho = 0.5, c_a^2 = 0$, we see that the errors decrease, because the dependencies decrease for a higher number of parallel servers. On the other hand, errors will be accumulated for a higher number of servers using the method for the stochastic merge subsequently and several times. We can find the consequences in the other results for $m = 3, \rho = 0.5$. We also see this accumulative error when we compare the errors for $m = 2, \rho = 0.9, c_b^2 = 0$ to the results for $m = 3, \rho = 0.9, c_b^2 = 0$.

We can summarize the results as follows:

1. For deterministic arrival and service processes the approximation causes high deviations.

2. The higher the interarrival time variability c_a^2 or the service time variability c_b^2, the better the approximation gets.

3. When the variabilities of both processes are high, the approximation results are very close to the exact results.

4. For an increasing number of parallel servers, the dependencies of the outgoing streams decrease and the approximation quality gets better, on the other hand errors caused by the merge of streams are accumulated, which has the opposite effect on the approximation quality.

Table 6.1.: interdeparture time variability for a system utilization of 0.5 and 0.9; comparison of analytical results and simulation results

m=2, ρ=0.5

	$c_b^2=0.0$			$c_b^2=0.1$			$c_b^2=0.5$			$c_b^2=0.9$		
c_d^2:	analy	sim	error %	analy	sim	error %	analy	sim	error %	analy	sim	error %
$c_a^2=0.0$	0.5827	0.0000	inf	0.5219	0.1999	161.07	0.6013	0.3550	69.36	0.6415	0.4221	52.00
0.1	0.5631	0.0997	464.81	0.5529	0.2984	85.31	0.6241	0.3826	63.14	0.6941	0.4404	57.62
0.5	0.5284	0.5025	5.15	0.5577	0.5950	-6.27	0.7381	0.6685	10.40	0.7161	0.6830	4.85
0.9	0.4549	0.5325	-14.56	0.6124	0.8126	-24.63	0.7686	0.8066	-4.71	0.9391	0.9121	2.96

m=2, ρ=0.9

	$c_b^2=0.0$			$c_b^2=0.1$			$c_b^2=0.5$			$c_b^2=0.9$		
c_d^2:	analy	sim	error %	analy	sim	error %	analy	sim	error %	analy	sim	error %
$c_a^2=0.0$	0.3400	0.0000	inf	0.4564	0.2814	62.22	0.7734	0.6864	12.68	1.0799	0.9762	10.63
0.1	0.3714	0.0999	271.75	0.4355	0.4439	-1.89	0.7618	0.7006	8.74	1.0373	0.9575	8.32
0.5	0.3970	0.2041	94.50	0.4883	0.5480	-10.88	0.7698	0.7761	-0.81	0.7687	0.7750	-0.81
0.9	0.4000	0.2799	42.93	0.4603	0.5517	-16.56	0.8028	0.8291	-3.17	1.0475	1.0454	0.20

m=3, ρ=0.5

	$c_b^2=0.0$			$c_b^2=0.1$			$c_b^2=0.5$			$c_b^2=0.9$		
c_d^2:	analy	sim	error %	analy	sim	error %	analy	sim	error %	analy	sim	error %
$c_a^2=0.0$	0.5602	0.0000	inf	0.6039	0.3161	91.05	0.6602	0.6189	6.67	0.7490	0.5420	38.19
0.1	0.6493	0.1250	419.45	0.6394	0.3877	64.94	0.7259	0.5930	22.41	0.7998	0.5617	42.40
0.5	0.6644	0.4999	32.91	0.6414	0.6878	-6.74	0.8048	0.7378	9.09	0.8012	0.7502	6.79
0.9	0.6396	0.8118	-21.21	0.6695	1.0298	-34.99	0.8181	0.8944	-8.53	0.9575	0.8848	8.21

m=3, ρ=0.9

	$c_b^2=0.0$			$c_b^2=0.1$			$c_b^2=0.5$			$c_b^2=0.9$		
c_d^2:	analy	sim	error %	analy	sim	error %	analy	sim	error %	analy	sim	error %
$c_a^2=0.0$	0.5017	0.0000	inf	0.6231	0.2404	159.21	0.9419	0.9127	3.19	1.0774	0.9110	18.26
0.1	0.5018	0.0546	819.36	0.5919	0.5970	-0.86	0.8807	0.8511	3.48	1.1582	0.9611	20.51
0.5	0.6360	0.4940	28.73	0.6649	0.9327	-28.72	0.8780	0.9480	-7.38	1.0305	1.0246	0.57
0.9	0.6163	0.8998	-31.50	0.6494	1.0050	-35.38	0.9121	0.9633	-5.31	1.1421	1.1394	0.23

7. Examples of Multi-Server Systems in Production and Service Networks

> Knowing is not enough; we must apply. Willing is not enough; we must do.
>
> *Johann Wolfgang von Goethe*

This chapter intents to give the reader an idea of the application possibilities of the presented calculation methods for multi-server queues. We first present an existing analysis of the material supply of a car assembly line and show the additional benefit, the multi-server queueing model can achieve. In a second case, we analyze the sterilization process in health establishments. We build a discrete time queueing network model and compare analytical results of a specific sterilization process to simulation results.

7.1. Material Supply of a Car Assembly Line

Many tasks in material handling systems are usually solved by accumulating orders and parts in batches and transporting them together. One reason is, that the transport in batches decreases the number of rides, and thus, saves labor cost. Another reason is that the standardization of large bins is much more advanced than that for small bins. This results in material handling and storage equipment for Euro-Pool pallets etc., such as forklift trucks and similar devices. A logical consequence is to use this standard equipment. In order to do this economically, items have to be transported in batches. Using forklift trucks might be a good solution for some tasks, but there are also cases, where it is very beneficial to be able to move small quantities independently and at the same time efficiently. Especially to supply a car assembly line, where space at the assembly stations is limited, small quantities of different parts are needed and a transport system, handling small container quantities is valuable.

Furmans, Schleyer and Schönung (2008) analyze the material supply of such a car assembly line, where small parts are supplied to the line in bins which have been provided from a picking area. The bins with the parts are stored in flow racks at the line, where the assembly worker can easily pick them up and use them for the assembly. The replenishment of the material is controlled by a Kanban-pull system. This means, that as soon as a bin gets empty, one bin quantity is picked in the picking area in order to replenish the material at the assembly line. The shelves at the line have to store enough bins in order to make sure that the material at the line is sufficient to meet the replenishment time with a predetermined probability (e.g. 95%). The material and therefore the necessary space can be minimized, when every single kanban is transmitted to the picking area immediately and every single full bin is transported to the line instantly. However, in practice milk-run trains are used and batches of bins and kanbans are transported. This reduces the transport cost but increases the necessary storage space in the shelves at the line, since the time used for batching and the waiting time at the picking area increases the replenishment time.

In order to realize a one-piece flow, Furmans et al. (2008) propose two new material flow devices:

SmartRack, a flow rack, which is able to realize e-Kanban easily and quickly using RFID-technology;

KARIS, a material handling system, that is able to move small quantities independently and autonomous.

7.1.1. Comparison of different material supply concepts

Furmans et al. (2008) analyze three different concepts for the material supply of the car assembly line, in order to quantify the effects of these new material handling devices. They calculate the replenishment time of material for each scenario using the discrete time toolbox presented in section 3.1. In the first scenario, the transport of full bins, respectively empty bins serving as container-kanban, is realized by a milk-run in regular intervals. In regular intervals, a milk-run vehicle starts a new tour, picking up the empty bins/kanbans from the line and simultaneously distributing full bins, which have been provided from the picking area (see figure 7.1, compare Furmans et al. 2008). The kanbans are then dropped off at the picking area. After picking, the full bins are loaded onto the milk-run train and subsequently distributed to their positions at the assembly line.

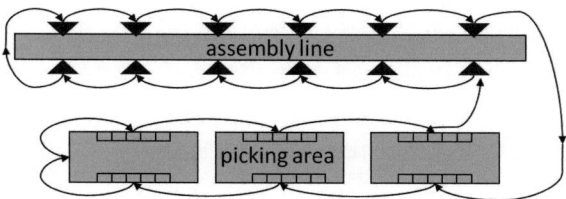

Figure 7.1.: Example of an assembly system

In order to study the effects of different milk-run frequencies on the replenishment time, the discrete time model illustrated in figure 7.2 is used (see Furmans et al. 2008). Since milk-runs are operated in regular intervals, the first step models the collecting process of kanbans by a batch building element for a specific time interval. These kanbans are then picked up and transported to the picking area. Assuming, that there is exactly one person picking, the picking process is modeled by a $G^x|G|1$-queue with batch arrivals. After picking, the transport of the full bins is again done in batches, occurring in fixed time intervals and after a transport time they arrive at their destination.

In a second scenario, the replenishment time can be reduced if the picking area is instantly notified that a bin is empty. This can be enabled

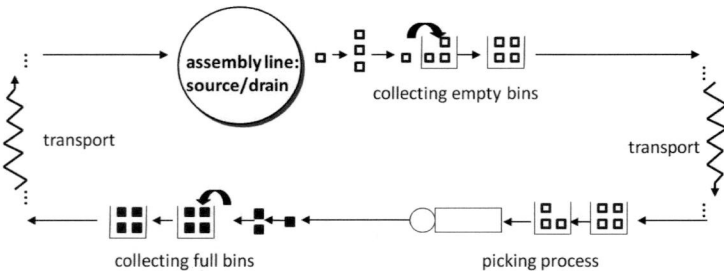

Figure 7.2.: Model of an assembly system

using RFID-technology. When a worker at the assembly line empties a bin and removes this bin from the flow rack, SmartRack recognizes this removal and an order is transferred to the picking area immediately. Thus, the collection of kanbans can be omitted as well as the transport of the empty bins from the assembly line to the picking area.

Finally, the best case is discussed. It is assumed that replenishment orders are transferred to the picking area instantly by means of RFID-technology and full bins are transported to the assembly line by single KARIS transport-elements instantly after picking. Since replenishment orders arrive singly at the picking area and full bins are transported singly to the assembly line, a one-piece flow (which means one-container flow) is realized. Therefore, the collection of full bins and the accordant batch building element can be removed from the model, too. Furmans et al. (2008) calculate the replenishment time in the one-piece flow as the sum of waiting time of orders in the picking area, picking time and the time for the transport from picking area to assembly line. In their model, the authors assume, that transport-elements are always available and full bins do not have to wait for transport.

For a numerical example, Furmans et al. (2008) analyze the replenishment time distribution for the previously presented scenarios. The discrete time distributions of the interarrival time of empty bins, respectively kanbans, at the assembly stations, the picking time for filling an empty bin, and the transport time for traveling between the assembly line and the picking area are given (see table A.1).

Using the presented models, Furmans et al. (2008) are able to calculate the sojourn time distributions for each of the successive processes of the underlying assembly system. This leads to the replenishment time distribution.

First, the replenishment time is analyzed according to the frequency of milk-runs resulting in different batch sizes (see figure 7.3, compare Furmans et al. 2008). Based on the replenishment time distributions, the mean replenishment time and some quantiles can be determined. The required time frame which guarantees the on-time replenishment of material with a given probability (e.g. 95% or 99%) decreases considerably with an increasing frequency of milk-runs (see table 7.1). The histogram of the replenishment time distribution of the RFID case clearly shows a further reduction of cycle time. The minimum possible replenishment time of the presented numerical case can be reached for the one-piece flow, also illustrated in figure 7.3 and table 7.1[1].

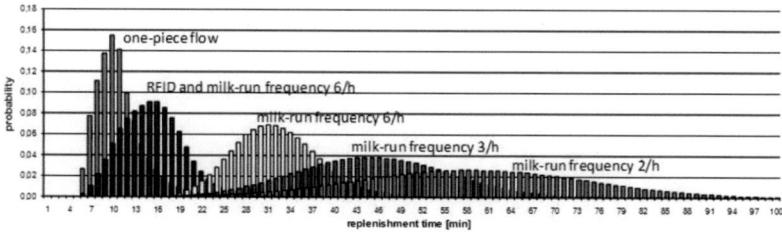

Figure 7.3.: Replenishment time distribution for different material supply scenarios

Once, the replenishment time distribution for a specific system configuration is known, the required stock of material at the assembly line can be determined (see Tempelmeier 2006a). Then, material flow planners are able to decide whether a higher milk-run frequency or the investment in new material handling devices, such as SmartRack and KARIS, is profitable.

---

[1]Note that in Furmans et al. (2008) a reading error for the quantiles of the best case appears. In our table we use the correct values.

	RFID one-piece flow	RFID milk-run frequency 6/h	milk-run frequency 6/h	milk-run frequency 3/h	milk-run frequency 2/h
expected value	10.03	14.51	31.21	44.49	57.87
95%-quantile	16	22	41	61	82
99%-quantile	21	27	47	68	92

Table 7.1.: The effect of increasing milk-run frequency, use of RFID and implementation of a one-piece flow on the replenishment time [min]

In their best case, Furmans, Schleyer and Schönung (2008) assume that an infinite number of KARIS transport elements are available to transfer full bins to the assembly line, immediately after picking. In practice, a planner will have to decide how many transport elements are necessary to cope with the material flow. Therefore, the discrete time methods for multi-server queueing systems are useful to determine the number of parallel transport elements needed for a one-piece flow, as demonstrated in the following section.

7.1.2. Determination of the number of parallel transport elements needed for a one-piece flow

Given is a certain number of parallel transport elements m. The transport of single bins can then be modeled as a $G|G|m$-queueing system (see figure 7.4). The material flow works as follows: As soon as a bin gets empty, the information is transmitted to the picking area electronically. One order picker is filling bins according to the singly arriving orders. If a transport element is available, a full bin can be carried to its destination at the assembly line. There, the empty bin is replaced by the full bin, and the transport element brings the empty bin to the picking area. Then, the transport element is ready to start another job. The service time of a transport job therefore includes the transport time for a round-trip. Using the presented model, the replenishment time distribution and its quantiles for different numbers of parallel transport units can be calculated. The replenishment time in this model is the sum of waiting time (picking), picking time, waiting time (transport) and transport time. Note that the replenishment time only includes the

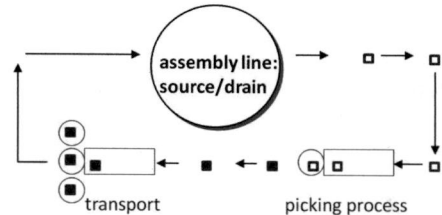

Figure 7.4.: Model of a material supply with one-piece flow

	number of parallel transport units				
	3	4	5	6	7
expected value	11.78	10.12	10.04	10.03	10.03
95%-quantile	20	16	16	16	16
99%-quantile	25	21	21	21	21

Table 7.2.: The effect of an increasing number of parallel transport units on the replenishment time

transport time from picking area to assembly line and excludes the way back. We implemented the queueing model in Java in order to calculate the replenishment time. Some numerical results for a different number of parallel transport-elements can be found in table 7.2. The material flow requires at least 3 parallel transport elements to meet the requested tours. Then, the replenishment time in 95% of the cases is smaller than or equal to 20 minutes, and the 99%-quantile is 25 minutes. Increasing the number of transport elements to 4 units, the quantiles decrease, but more than 4 parallel elements do not have any further effect on the quantiles. Only the expected value of the replenishment time decreases slightly. Note, that the calculation of the replenishment time distribution only takes a few milliseconds for 3 parallel transport units, and increases to some seconds for 7 units. In contrast, one run of a corresponding simulation model takes several minutes to hours, depending on the desired accuracy of the results.

Looking at the waiting time distribution of transport jobs (see table A.2), we can have a more detailed impression of the benefits further transport elements can achieve. We can see, that for more than 3 transport elements, only a few transport jobs will have to wait. For 7 or more transport elements, every full bin can be transported immediately without waiting.

This numerical results can now help a material flow planner to compare cost and benefit of each scenario and also for each additional transport element.

7.2. Sterilization Processes in Health Establishments

Health establishments are faced with growing health expenses, resulting from the increase of the medical acts cost, while the society tries more and more to reduce health expenses in order to guarantee its social welfare. We can thus see a well known problem in the context of production systems: making a system more efficient, in order to limit the increase of expenses.

In health establishments, the sterilization of surgical and exploration instruments plays a key role in the fight against infections. In Di Mascolo et al. (2006), the authors studied and modeled a particular sterilization service. Then, they conducted a survey, they sent to health establishments in the Rhône-Alpes region in France (see Reymondon et al. 2008), motivated by the wish to know if the services are homogeneous. They found out, that the sterilization processes are following a relatively similar structure. Using the information provided by the survey, Di Mascolo et al. (2009) obtained a generic model structure of a centralized sterilization service. This structure was used to propose a queueing network model for a generic sterilization service, and to build a simulation model. The aim was to compare the performance of the participants of the survey to identify the most successful services. The model should also be used for a dimensioning of the resources (e.g. number of washers, autoclaves, staff,...) in a sterilization service. Using existing results on discrete queueing models, the queueing network model could not be analyzed yet. One missing element was a multi-server queue. Thus, the simulation model was used to provide some performance parameters (see Di Mascolo et al. 2009), which could not be obtained with the survey, and are useful for the comparison of health establishments performance. As we are now able to analyze a multi-server queue, we can model a queueing network, representing a sterilization process. We will first introduce the reader to the generic structure of the sterilization process. Then, we present a queueing network model and some performance figures of a specific health establishment, obtained by this model.

7.2.1. Generic Structure of a Sterilization Process

In a sterilization process, reusable medical devices are re-injected in the process after their use in the operation room. When we integrate the use step, the sterilization process becomes a sterilization loop as seen in figure 7.5 (Di Mascolo et al. 2006). The pre-disinfection step is

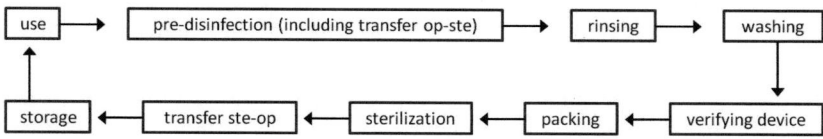

Figure 7.5.: Sterilization loop

done directly after the use in the operation room. The medical devices are placed in a disinfectant liquid to decrease the population of micro-organisms present on the soiled equipment, in order to protect the staff during the manipulation and to facilitate the later washing. During the pre-disinfection step, the used medical devices (MD) are transferred from the operation rooms down to the sterilization area. At the sterilization area the MD are rinsed, but this step does not always exist and can be included in the washing step. The MD are washed in machines to eliminate stains to obtain a clean medical device. After the washing, the MD are checked to ensure that no deterioration may affect their security and functionality. After washing, the MD are packed to constitute a barrier against micro-organisms. In the sterilization step, the MD are placed in an autoclave where they are treated with saturated steam. The transfer ste-op corresponds to the transfer of sterile MD from the sterilization service up to the storage area close to the operating rooms.

7.2.2. Queueing Network Model of a Sterilization Process

We present here the structure of a model enabling to represent the production of sterile medical devices, excluding the use of MD in operating rooms and the storage before the use. Our model thus contains the main steps of the sterilization loop, that take place in the sterilization area, namely pre-disinfection, rinsing, washing, packing and sterilization. The verification after the washing is included in the packing step.

At the input of the model, the entities arriving at the pre-disinfection step are called "operations" and represent a batch of containers and bags of medical devices, used for one given surgical operation. The choice of this entity enables to ensure that all containers and bags that served for a given surgical operation are washed together, in the same washer. The operations are transferred to the sterilization area in batches. Then they are rinsed one by one. The operations are then washed in batches in one of the parallel washers. After the washing step, the "operations" are divided into containers and into bags, whose quantities depend on the operation. Containers and bags are packed at separate stations. After packing, the containers and bags are merged. In the sterilization step, the containers and bags are then treated as the same kind of unit ("equivalent containers"), while a certain number of bags equates one container unit. The sterilization step is processed in batches of "equivalent containers", with a fixed batch size, in one of the parallel autoclaves.

In order to improve the performance of the system, different scenarios for the transfer of MD to the sterilization area can be analyzed. In the following model, we choose a transfer of MD to the sterilization area in regular time intervals in order to get an even flow of parts (see Di Mascolo et al. 2006). Other scenarios would be e.g. a transport when a certain number of operations is collected (capacity rule). In practice, the transport is normally not following a certain rule. This unsteadiness can cause a duration of pre-disinfection that does not lead to the desired effect on the material.

Using the existing discrete time methods (chapter 3) and the methods for the analysis of multi-server queues, we can model the sterilization process as follows (see figure 7.6).

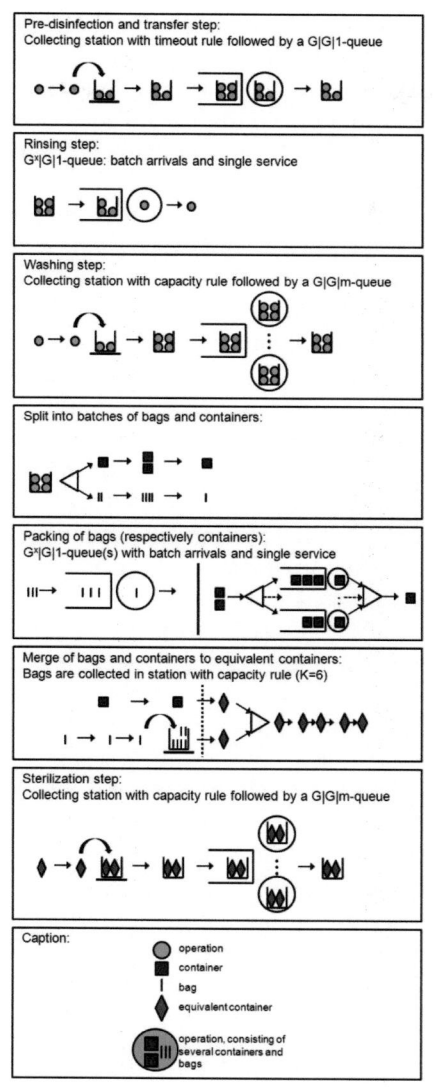

Figure 7.6.: Queueing network model of a sterilization process

71

When the MD are transferred to the sterilization area by an elevator in regular intervals, the pre-disinfection and transfer step can be modeled by a collecting station with a timeout rule, followed by a $G|G|1$-queue. The customers arrive in single units (operations) and build batches of variable size. Then one batch of operations is treated as a customer in the transfer-queue. At the rinsing step, a batch of operations arrives, and single operations are served. So, we can model this step by a $G^x|G|1$-queue with batch arrivals and single service. To fill the washer, a certain number of operations is collected. As the size of the operations is variable, we can not fix the number of operations that fit the capacity of a washer. We thus make an assumption. We take the average batch size of operations that can be loaded to one washer and then always collect a batch with this fixed capacity K. We thus simplify our model by considering that a fixed number of operations are processed by the washers and the autoclaves, which is the same assumption as in the simulation model (Di Mascolo, Gouin and Ngo Cong 2006). So, the washing step can be modeled as a collecting station with capacity rule followed by a $G|G|m$-queue, where each arriving batch of operations is treated as a single customer. After washing, the batches of operations are split into containers and into bags. The batches of containers, respectively bags, are arriving at the according packing step, where they are processed in single units. Normally, there are several parallel working stations for the packing of containers. This step can thus be represented by a $G^x|G|m$-queue. As this queueing element is not solved yet, we approximate the step by m parallel $G^x|G|1$-queues, with m separate waiting rooms. The arriving containers then have to be split in front of these parallel stations and merged after the packing step. We therefore use the stochastic split and merging elements. After the packing step, the streams of containers and bags are consolidated by a stochastic merging element. As the streams contain different kind of units, we first transform a batch of 6 bags to one "equivalent container" that equates one container size (compare Ngo Cong 2009). This step is represented by a collecting station with capacity rule. Then, the streams of "equivalent containers" are merged. Before the sterilization in the autoclaves starts, a certain number of "equivalent containers" is collected. This can also be modeled by a collecting station with capacity rule similar to the washing step. We then have a $G|G|m$-queue, representing the sterilization step, where the arriving customers are batches of containers and one batch is treated as one customer. The single steps of the sterilization process and their corresponding models are summarized in figure 7.6. Note that some other rules could be used to model the building of batches at the washing or sterilization steps. Some examples of other rules can be found in (Di Mascolo, Gouin and Ngo Cong 2009) or in (Özden, Berbig, Matzka, Furmans and Di Mascolo 2010). We choose here the same as the one taken in the simulation model, which serves as a reference for the results.

7.2.3. Numerical Results

We now use the presented queueing model to analyze some performance figures of a particular sterilization process. In a previous work, Di Mascolo, Gouin and Ngo Cong (2006) analyzed a specific health establishment (Centre Hospitalier Privé Saint Martin de Caen) via simulation. We use the same input data (see Ngo Cong 2009) to compare our queueing network model to the accordant simulation model. In both models, the operations are collected after use within a fixed time interval $t_{out} = 30$ minutes. The regarded sterilization process consists of 4 parallel washers and 3 autoclaves. In the packing area, 4 persons are handling the containers and one person is responsible for the packing of bags. As we can not reach the same level of detail in the discrete time model as in the simulation model, we have to make some assumptions:

- In the interarrival time distribution of operations at the pre-disinfection step, some values are higher than the collecting time t_{out}. As the according model for the collecting process (Schleyer 2007) can just be used for interarrival times smaller than t_{out}, we cut all values exceeding t_{out} and put the remaining probability mass on the last element of the distribution.

- We assume that the washers are loaded with a fixed number of operations, each with the same size. In fact, the capacity of these machines depends on the variable size of the operations and fits not always the same number of operations.

- For the packing of containers, we consider a constant number of workers. In reality, the number varies over the course of a day.

73

- The packing of containers is approximated by m parallel $G^x|G|1$-queues. We assume that the arriving containers are equally distributed to the m service stations.

As mentioned above, we implemented the multi-server queuing models in Java and linked them to existing models (see Schleyer 2007). The execution time of the queueing network model takes only a few seconds. In contrast, the simulation model (built in the simulation software pro-model) in (Di Mascolo, Gouin and Ngo Cong 2006) requires several minutes or even hours, depending on the desired accuracy of the results and thus the required throughput. From the queueing network model, we obtain two important performance figures that can be compared to the simulation results. One of them is the average duration of the pre-disinfection step. The ideal duration of pre-disinfection, to guarantee an optimal impact of the disinfection liquid to the medical devices, is about 15 minutes. On the other hand, the sojourn time in the liquid should not exceed 50 minutes, because the disinfection product attacks the material, and thus causes a premature ageing. We thus want to know the average pre-disinfection time as well as the percentage of operations that stay in the liquid more than 50 minutes. Compared to simulation, our queueing network model obtains quite good results for the average pre-disinfection time and the percentage of medical devices, that stay in the liquid more than 50 minutes. The values obtained by the queueing model are marginally higher than the appropriate simulation results. This effect is caused by the fact that our interarrival time distribution is fitted to the collecting time t_{out}, and thus, the mean value of the interarrival time is smaller. A smaller interarrival time is equivalent to a higher utilization of the system and longer waiting and sojourn times. We can conclude, that the queueing network model is capable to analyze the influence of the transfer interval between operation area and sterilization area. To obtain more accurate values, a modified method for the collecting under timeout rule would be useful. This method should be able to handle interarrival time distributions with values higher than t_{out}.

A second important parameter is the sojourn time of medical devices through the sterilization process before they are ready to be used again. From the survey in the Rhône-Alpes region (see Reymondon et al. 2008), we know that many health establishments are not able to estimate the

	simu	analysis	deviation [%]	deviation abs.
av. pre-dis. time [min]	29.40	30.50	3.74	1.10
P(pre-dis. ≤ 50 min) [%]	92.90	91.46	-1.55	-1.44
av. sojourn time [min]	454.00	345.03	-24.00	-108.97

Table 7.3.: performance figures obtained by analysis and simulation

duration of their sterilization process. The average sojourn time given in table 7.3 can give them an idea of the cycle time of medical devices. This figure also allows us to compare different loading policies for the washers and autoclaves and the influence of the number of parallel machines to the sojourn time. As we see in table 7.3, the average sojourn time calculated by our queueing model is 24% smaller than the value obtained via simulation. The reason for this high deviation is supposed to be the approximation of unsolved queueing models by known modeling elements and the assumption of independent nodes in the queueing network we made in order to be able to calculate the distribution of the sojourn time by convolution.

8. Conclusion and Outlook

Genius begins great works,
labor alone finishes them.

Joseph Joubert

Motivated by the advantages of discrete time modeling for the analysis of material flow and service systems, a lot of important discrete time queueing models with generally distributed processes had been developed so far but there was one central element missing in this "toolbox": the discrete time multi-server queueing system. As parallel servers are existent in material flow systems in many different forms, an appropriate model is valuable to increase the accuracy of several models of material flow systems. Because of its numerous application possibilities, the multi-server queue is well researched in continuous time domain. However, an adequate general discrete time queueing model for the analysis of material flow systems, dealing with parallel servers, has not been researched yet. This motivated the development of discrete time methods for the analysis of multi-server queues.

In the current thesis, we presented new analytical methods for the performance evaluation of $G|G|m$-queues. The distributions for the number of customers in the system, the waiting time and the interdeparture time were calculated subsequently.

First, we presented a discrete time method for the computation of the distribution of the number of customers at an arrival instant. We calculated the probabilities to see a certain system state at an arrival instant, using a discrete homogenous Markov chain that we embedded at the arrival instants. In the presented method, the first step is to calculate the transition probabilities from a state at the arrival of a certain customer to the system state at the arrival of the succeeding customer. Therefore, we distinguished between three different cases according to the utilization of the system at the two succeeding arrival moments. The presented method is exact within an ϵ-environment.

Based on the distribution of the number of customers at the arrival instant, we presented an exact method for the calculation of the waiting time distribution in a multi-server queueing system. Iterating over all possible system states and the possible service time values for each customer, we allocated the waiting customers to the servers according to the rule First-come, first-served. We then were able to determine the waiting time for each iteration step and the corresponding probability. Knowing the waiting time distribution of an arbitrary customer as well as the service time distribution, we were also able to calculate the sojourn time distribution.

With the distribution of the number of customers in the system and the waiting time, we are able to analyze the multi-server queue separately. For the analysis of complete material flow networks, we need an interface to connect the multi-server queue to other nodes of a network, like e.g. queues or merges or splits. This interface is given by the interdeparture time distribution of customers that leave the multi-server queue after their service. Therefore, we presented an approximation method for the calculation of the interdeparture time distribution. For the calculation of the interdeparture time distribution of a multi-server queue, we first analyzed the output stream of a single server and then merged the single streams using a stochastic merging element. To determine the interdeparture time of a single server, we started with the analysis of the idle time, that occurs, when a customer leaves the system and there is no customer present in the waiting room to start his process. In contrast to a single-server queue, in a multi-server queue the idle time does not automatically end, when a new customer enters the system. Taking this information into account, we were able to calculate the idle time distribution, and thus, the interdeparture time distribution of a single server, again using an iterative method. The presented method for the calculation of the interdeparture time distribution of a single server is exact, but as the stochastic merge is an approximation, the distribution of the outgoing stream of the multi-server queue is not exact. The stochastic merge is approximating a point process by a renewal process. Therefore, the approximation causes errors for deterministic interarrival and service processes and is accurate for high variabilities.

Applying the presented methods to model the material supply of an assembly line and the sterilization process of medical devices in health establishments, we were able to show that the discrete time methods are well suited to analyze material flow systems. Analyzing the sterilization process, we derived accurate results for some performance figures, but obtained high deviations for others. The reason for this impreciseness on the one hand is the lack of appropriate analytical models. On the other hand, dependencies in a queueing network can not be taken into account with the existing discrete time queueing model elements. This is an aspect that should be regarded in further studies. We also identified the need for the following model elements which can be addressed by further research:

- the multi-server queue with batch arrivals and single service ($G^x|G|m$-queue)

- the multi-server queue with batch arrivals and batch service ($G^x|G^x|m$-queue)

- the multi-server queue with inhomogeneous servers which is often found when a machine pool is extended by a new and faster server

- the existing method for the analysis of batch building processes under timeout rule (Schleyer 2007) is restricted to interarrival time values smaller than the collecting time t_{out}. An extension of the method for values higher than t_{out} would be valuable in order to avoid approximations of the interarrival time distribution.

- an analogue extension can be useful for the batch building process under capacity rule (Schleyer 2007), that currently is restricted to values smaller than the given batch building capacity K.

With the presented methods for the analysis of multi-server queueing systems, we are able to make a contribution to an efficient and accurate calculation of performance figures of material flow and service systems. Nevertheless, further research in the area of discrete time queueing analysis and the extension of the discrete time "toolbox" is required.

Glossary of Notation

A	random variable describing the interarrival time process
a_α	interarrival time distribution
α	interarrival time
B	random variable describing the service process
b_β	service time distribution
β	service time
β^q	service time of customer q, that is waiting in the queue
c_a^2	variability of the interarrival time
c_b^2	variability of the service time
c_d^2	variability of the interdeparture time
CDF	cumulative distribution function
d_δ	interdeparture time distribution
δ	interdeparture time
$\vec{d}_{single}$	interdeparture time distribution vector of a single server
$\vec{d}$	interdeparture time distribution vector of the multi-server system
k_κ	sojourn time distribution
κ	sojourn time
i	system state at the arrival of customer c
$i.i.d.$	independent and identical distributed
j	system state at the arrival of customer $c+1$, which is the successor of customer c
m	number of parallel servers
MD	medical device
n_η	distribution of the number of customers in the system at the arrival instant of an arbitrary customer
η	number of customers in the system at the arrival instant of an arbitrary customer

η^c	number of customers in the system at the arrival instant of customer c
η^{c+1}	number of customers in the system at the arrival instant of customer $c+1$
Ξ	number of busy servers with residual service time of zero
P	transition matrix with elements p_{ij}
p_{ij}	probability for a transition from system state i to state j
$p_{idle \rightarrow busy}$	probability that an idle server gets busy at the arrival of the next customer
$p_{idle \rightarrow idle}$	probability that an idle server stays idle at the arrival of the next customer, because the customer entered another server
pmf	probability mass function
π	system state at the arrival moment of an arbitrary customer
q	label for the customers waiting in the queue
r_s	residual service time of server s
$\vec{r}$	vector with residual service times of all servers
r_s^c	residual service time of server s at the arrival moment of customer c
$\vec{r}^c$	vector with residual service times of all servers at the arrival moment of customer c
r_s^{c+1}	residual service time of server s at the arrival moment of customer $c+1$
$\vec{r}^{c+1}$	vector with residual service times of all servers at the arrival moment of customer $c+1$
ρ	utilization of the queueing system
s	serial number of a certain server
scv	squared coefficient of variation
$\vec{t}$	vector with service times of all customers waiting in the queue
τ	idle time of a server
τ^I	initial idle time of a server

$U^{\alpha,\vec{r}^c,\vec{r}^{c+1}}$	random variable describing the number of customers the system can serve within a certain interarrival time interval α, starting from a system state with a certain constellation of residual service times $\vec{r}^c$ and ending up in a constellation $\vec{r}^{c+1}$
$\vec{u}^{\alpha,\vec{r}^c,\vec{r}^{c+1}}$	probability vector for the number of customers the system can serve within a certain interarrival time interval α, starting from a system state with a certain constellation of residual service times $\vec{r}^c$ and ending up in a constellation $\vec{r}^{c+1}$
$\vec{u}^{\alpha,r_s^c,r_s^{c+1}}$	probability vector for the number of customers server s can serve within a certain interarrival time interval α, starting from a residual service time r_s^c and ending up in a constellation r_s^{c+1}
$u_v^{\alpha,\vec{r}^c,\vec{r}^{c+1}}$	probability that the system can serve exactly v customers within a certain interarrival time interval α, starting from a system state with a certain constellation of residual service times $\vec{r}^c$ and ending up in a constellation $\vec{r}^{c+1}$
v	number of customers the system serves within a certain interarrival time interval α, starting from a system state with a certain constellation of residual service times $\vec{r}^c$ and ending up in a constellation $\vec{r}^{c+1}$
w_ω	waiting time distribution
ω	waiting time
wta_s	working time account of server s
X	random variable describing the system state at the arrival of an arbitrary customer
x_π	distribution of the number of customers in the system at the arrival moment of an arbitrary customer
$\vec{y}$	idle time distribution vector
$y_{\tau^I}^I$	distribution of the initial idle time
$y_{\tau^{I*}}^{I*}$	distribution of the initial idle time including idle times of zero
z	server with the lowest working time account
$\otimes$	convolution operator

Bibliography

Ackroyd, M. H. (1980). Computing the Waiting Time Distribution for the G/G/1 Queue by Signal Processing Methods. *IEEE Trans. Commun. COM-28*, p. 52–58.

Alfa, A. S. (2002). Discrete Time Queues and Matrix-Analytic Methods. *Sociedad de Estadística e Investigacíon Operativa 10*(2), p. 147–210.

Arnold, D. and K. Furmans (2009). *Materialfluss in Logistiksystemen* (5. Edition). Springer.

Bertsimas, D. (1988). An Exact FCFS Waiting Time Analysis for a General Class of G/G/s Queueing Systems. *Queueing Systems 3*, p. 305–320.

Björklund, M. and A. Elldin (1964). A Practical Method of Calculation for Certain Types of Complex Common Control Cystems. *Ericsson Technics 20*, p. 2–75.

Bolch, G. (2006). *Queueing networks and Markov chains: modeling and performance evaluation with computer science applications.* Hoboken, NJ: Wiley.

Boxma, O. J., J. W. Cohen and N. Huffels (1979). Approximations of the Mean Waiting Time in an M/G/s Queueing System. *Operations Research 27*(6), p. 1115–1127.

Breuer, L. (2003). Transient and Stationary Distributions for the GI/G/k Queue with Lebesgue-Dominated Inter-Arrival Time Distribution. *Queueing Syst. Theory Appl. 45*(1), p. 47–57.

Brumelle, S. L. (1971). Some Inequalities for Parallel-Server Queues. *Operations Research 19*(2), p. 402–413.

Burman, D. and D. Smith (1983). A Light-Traffic Theorem for Multi-Server Queues. *Mathematics of Operations Research 8*, p. 15–25.

Buzacott, J. A. and J. G. Shanthikumar (1993). *Stochastic models of manufacturing systems.* Englewood Cliffs, N.J: Prentice Hall.

Choi, D., N. Kim and K. Chae (2005). A Two-Moment Approximation for the GI/G/c Queue with Finite Capacity. *INFORMS Journal on Computing 17*(1), p. 75–81.

Daduna, H. (2001). *Queueing Networks with Discrete Time Scale - Explicit Expressions for the Steady State Behavior of Discrete Time Stochastic Networks*, Volume 2046 of *Lecture Notes in Computer Science*. Springer.

Di Mascolo, M., A. Gouin and K. Ngo Cong (2006). Organization of the production of sterile medical devices. In: *Proceedings of the 12th IFAC Symposium on Information Control Problems in Manufacturing - INCOM 2006: Saint Etienne, France*.

Di Mascolo, M., A. Gouin and K. Ngo Cong (2009). A generic model for the performance evaluation of centralized sterilization services. In: *Proccedings of the conference on Stochastic Models of Manufacturing and Service Operations, SMMSO 2009: Lecce, Italy*.

Furmans, K. (2000). *Bedientheoretische Methoden als Hilfsmittel der Materialfluplanung*. Institut für Fördertechnik und Logistiksysteme der Universität Karlsruhe (TH).

Furmans, K. (2004). A Framework of Stochastic Finite elements for Models of material Handling systems. In: *8. International Material Handling Research Colloquium, Graz*.

Furmans, K., M. Schleyer and F. Schönung (2008). A Case for Material Handling Systems, Specialized on Handling Small Quantities. *10. International Material Handling Research Colloquium, Dortmund*.

Furmans, K. and A. Zillus (1996). Modeling independent production buffers in discrete time queueing networks. In: *Proceedings of CIMAT '96, Grenoble*, p. 275–280.

Gnedenko, B. W. and D. König (1983). *Handbuch der Bedientheorie I*, Volume 1. Akademie-Verlag, Berlin.

Grassmann, W. K. and J. L. Jain (1989, Jan. - Feb.). Numerical Solutions of the Waiting Time Distribution and Idle Time Distribution of the Arithmetic GI/G/1 Queue. *Operations Research 37*(1), p. 141–150.

Greiling, M. (1997). *Verbesserung der Produktionslogistik durch Losgrenharmonisierung - Ein bedientheoretischer Ansatz*. Dissertation, Universitt Karlsruhe, Institut fr Frdertechnik und Logistiksysteme.

Halfin, S. and W. Whitt (1981). Heavy-Traffic Limits for Queues with Many Exponential Servers. *Operations Research 29*(3), p. 567–588.

Haßlinger, G. (1995). A Polynomial Factorization Approach to the Discrete Time GI/G/1/(N) Queue Size Distribution. *Perform. Eval. 23*(3), p. 217–240.

Hopp, W. J. and M. J. Spearman (2001). *Factory physics* (2. ed. Edition). Boston [u.a.]: Irwin/McGraw-Hill.

Hübner, F. and P. Tran-Gia (1995). Discrete-time analysis of cell spacing in ATM systems. *Telecommunications Systems Vol. 3, pp. 379-395.*

Jain, J. L. and W. K. Grassmann (1988). Numerical solution for the departure process from the GI/G/1 queue. *Computers & OR 15*(3), p. 293–296.

Kim, N., K. Chae and M. Chaudhry (2004). An Invariance Relation and a Unified Method to Derive Stationary Queue-Length Distributions. *Operations Research 52*(5), p. 756–764.

Kimura, T. (1986). A Two-Moment Approximation for the Mean Waiting Time in the GI/G/s Queue. *Management Science 32*(6), p. 751–763.

Kimura, T. (1994). Approximations for Multi-Server Queues : System Interpolations. *Queueing Systems 17*, p. 347–382.

Kingman, J. (1962). Some Inequalities for the Queue GI/G/1. *Biometrika 49*, p. 315–324.

Kingman, J. F. C. (1970). Inequalities in the Theory of Queues. *Journal of the Royal Statistical Society 32*(1), p. 102–110.

Kleinrock, L. (1976). *Queueing Systems, Vol. 2: Computer Applications.* Wiley, New York.

Kleinrock, L. and R. Gail (1996). *Queueing systems : problems and solutions.* A Wiley-Interscience publication. New York, NY [u.a.]: Wiley.

Lee, A. M. and P. A. Longton (1959). Queueing Processes Associated with Airline Passenger Check-in. *OR 10*(1), p. 56–71.

Lippolt, C. (2003). *Spielzeiten in Hochregallagern mit doppeltiefer Lagerung.* Dissertation, Universität Karlsruhe, Institut für Fördertechnik und Logistiksysteme.

Little, J. D. C. (1961). A Proof for the Queuing Formula: $L = \lambda \cdot W$.

Operations Research 9(3), p. 383–387.

Matzka, J., M. Di Mascolo and K. Furmans (Online first: Sept. 2009). Buffer Sizing of a Heijunka Kanban System. *Journal of Intelligent Manufacturing* (DOI: 10.1007/s10845-009-0317-3).

Neuts, M. (1981). *Matrix-Geometric Solutions in Stochastic Models.* John Hopkins University Press.

Ngo Cong, K. (2009). *Etude et Amelioration de l'Organisation de la Production de Dispositifs Medicaux Steriles.* Thèse de doctorat, Université Joseph Fourier.

Özden, E., D. Berbig, J. Matzka, K. Furmans and M. Di Mascolo (2010). Discrete Time Analysis of Batch Building Processes with the Capacitated Timeout Rule. *Annals of Operations Research.* under review.

Özden, E. and K. Furmans (2010). Analysis of the discrete-time $G^X|G^{[L,K]}|1$-queue. In: *24th European Conference on Operational Research, Lisbon.*

Page, E. (1972). *Queueing Theory in OR.* Butterworth.

Puhalskii, A. A. and M. I. Reiman (2000). The Multiclass GI/PH/N Queue in the Halfin-Whitt Regime. *Advances in Applied Probability 32*(2), p. 564–595.

Rall, B. (1998). *Analyse und Dimensionierung von Materialflusystemen mittels geschlossener Warteschlangennetze.* Dissertation, Universität Karlsruhe, Institut für Fördertechnik und Logistiksysteme.

Reymondon, F., M. Di Mascolo, A. Gouin, B. Pellet and E. Marcon (2008). Etat des lieux des pratiques de stérilisation hospitalière en Rhône-Alpes. In: *Proceedings of GISEH 2008: 4ème conférence francophone en Gestion et Ingenierie des SystèmEs Hospitaliers.*

Sakasegawa, H. (1977). An Approximation Formula. *Ann. Inst. Statist. Math. 29*(2), p. 67–75.

Schleyer, M. (2007). *Discrete time analysis of batch processes in material flow systems.* Dissertation, Universität Karlsruhe, Institut für Fördertechnik und Logistiksysteme.

Schleyer, M. and K. Furmans (2007a). An analytical method for the calculation of the waiting time distribution of a discrete time G/G/1-queueing system with batch arrivals. *OR Spectrum 29*(4), p. 745–763.

Schleyer, M. and K. Furmans (2007b). A framework of Stochastic Finite Elements for Models of Manufacturing Systems. In: *Proceedings of the Sixth International Conference on Analysis of Manufacturing Systems, Eindhoven, The Netherlands.*

Schleyer, M., K. Furmans and M. Di Mascolo (2007). Modeling of manufacturing systems - A discrete time approach. In: *Proceedings of the IAR Annual Meeting, Grenoble.*

Seelen, L. and H. Tijms (1984). Approximations for the Conditional Waiting Times in the GI/G/c Queue. *Operations Research Letters 3*(4), p. 183–190.

Tempelmeier, H. (2006a). *Inventory Management in Supply-Networks – Problems, Models, Solutions.* Norderstedt: Books on Demand.

Tempelmeier, H. (2006b). Supply chain inventory optimization with two customer classes in discrete time. *European Journal of Operational Research 174*(1), p. 600–621.

Tempelmeier, H. and L. Fischer (2009). Approximation of the probability distribution of the customer waiting time under an (r, s, q) inventory policy in discrete time. *International Journal of Production Research.*

Tran-Gia, P. (1996). *Analytische Leistungsbewertung verteilter Systeme: eine Einführung; mit Übungsaufgaben.* Springer-Lehrbuch. Berlin: Springer.

Walrand, J. (1983). A discrete-time queueing network. *Journal of Applied Probability*, p. 903–909.

Whitt, W. (1983). Queueing network analyser. *The Bell System Technical Journal 62*(9), p. 2779–2815.

Whitt, W. (2005). Heavy-Traffic Limits for the $G|H_2^*|n|m$ Queue. *Math. Oper. Res. 30*(1), p. 1–27.

Wolff, R. W. (1989). *Stochastic modeling and the theory of queues.* Prentice Hall international series in industrial and systems engineering. Englewood Cliffs, NJ: Prentice Hall.

Zillus, A. (2003). *Untersuchung der Wartezeit von Kundenaufträgen in der Supply Chain.* Dissertation, Universität Karlsruhe, Institut für Fördertechnik und Logistiksysteme.

A. Appendix

minutes	interarrival time distr. of empty bins	picking time distr.	transport time distr.
0	0	0	0
1	0.2	0	0
2	0.175	0.3	0
3	0.15	0.5	0.2
4	0.1	0.15	0.2
5	0.1	0.05	0.2
6	0.1		0.2
7	0.075		0.2
8	0.05		
9	0.05		

Table A.1.: Given distributions of the interarrival time, picking time and transport time

waiting time [min]	number of parallel transport units				
	3	4	5	6	7
0	0.5352731	0.9543493	0.9985374	0.9999922	1.0000000
1	0.1004698	0.0267297	0.0011844	0.0000077	0.0000000
2	0.0921221	0.0126431	0.0002476	0.0000001	
3	0.0768815	0.0046909	0.0000298		
4	0.0588498	0.0012839	0.0000008		
5	0.0422160	0.0002522			
6	0.0292675	0.0000425			
7	0.0200371	0.0000063			
8	0.0138653	0.0000019			
9	0.0096234	0.0000002			
10	0.0066590				
11	0.0046271				
12	0.0031765				
⋮	⋮				
35	0.0000001				

Table A.2.: Transport waiting time distribution depending on the number of parallel transport units